NOISE POLLUTION is indispensable not only for the concerned citizen, but for all those who can, and must, take immediate and effective action in our unquiet crisis: Urban Planners, Architects, Hospital Administrators, Public Health Officials, Transportation Executives, Lawyers, Realtors, Sound Engineers, manufacturers of transportation equipment and household appliances, and community leaders. It is a vital resource in dealing with the noise crisis that is destroying pleasure, lowering work performance, eroding health, causing physical injury, and even challenging basic human survival.

CLIFFORD R. BRAGDON, an Environmental Specialist with the Bio-Acoustical Division of the U. S. Army Environmental Hygiene Agency, is Chairman of the American National Standards Institute's S3-50 Committee on Measurement and Evaluation of Outdoor Community Noise; he is also Associate Professor of City Planning at the Georgia Institute of Technology. Formerly a Research Associate at the Institute of Environmental Studies at the University of Pennsylvania, he holds a Master's in Urban Planning from Michigan State University and a Doctorate in City Planning from the University of Pennsylvania.

Noise Pollution

The Unquiet Crisis

Noise Pollution

The Unquiet Crisis

CLIFFORD R. BRAGDON

University of Pennsylvania Press

PHILADELPHIA

Designed by Howard King
Printed in the United States of America

The gracefulness of the Dorris does not stop with the beauty of its lines—it is expressed as noticeably in the noiseless, sweeping ease with which it moves—the way it floats up to a stop—the way it glides away from the standstill—the low murmur of the motor—the soundlessness of the gears in changing. All these things endow the Dorris with that grace which is inherent in well-bred people and things.

Advertisement prepared by D'Arcy Advertising Company for Dorris Motor Car Co., *St. Louis Post Dispatch* July 2, 1916. (Dudley A. Bragdon, copywriter; pers. comm.)

Contents

List of Figures

List of Tables

Preface

Urban noise pollution has rapidly grown to be a major environmental problem. A threat to physical and psychological well-being, the sounds of our technology follow us through our working, leisure, and sleeping hours. Although some research has been done on the effects of noise, and a few federal, community, and private groups have tried to work toward reducing noise pollution, this unquiet crisis has received little attention in the United States, less attention, in fact, than it has in Europe.

Noise as an Environmental Concern

This syndrome of modern society deserves further recognition as a serious environmental liability for several reasons. First, noise is pervasive. The chance of gaining refuge from noise is disappearing rapidly. Garbage trucks and early hour jet flights signal the beginning of each day. Often these noises start the day prematurely. In our homes, sounds of the neighbors' children, pets, and television pierce the morning silence. Outside, the wheels of urban activity are beginning to turn. The noise level grows with the quickening pace of morning activity.

NOISE IS ALL AROUND US

On our way to work, the din of traffic, building construction, and street repair continually assault us. The work environment is another auditory experience. Noise, long associated only with factory operations, is a problem afflicting most offices. Modern office design with expansive glass areas, open work spaces, insufficient partitioning, contemporary furniture that does not absorb sound, and mechanical office equipment can create an acoustic chaos. Home offers little refuge when the urban dweller returns in the evening. Heating-ventilating systems, plumbing, and home appliances hiss, chug, hum, swish, and grind indoors while the walls only partially subdue the roar of activity from outside. We are aware in mind and body that noise is an ever-present part of contemporary urban living.

NOISE SOURCES ARE INCREASING

Pervasiveness is just one characteristic of the problem. Noise sources are multiplying rapidly. The industrial and technological development of urban society is producing an increasing number of devices with higher and higher noise outputs. Aircraft, automobiles, trucks, motorcycles, construction equipment, household appliances, lawn-mowers, and air conditioners all contribute to a noisier environment. The list grows endlessly: garbage disposals, blenders, snow blowers, snowmobiles, studded snow tires, and so on. Equally important is the increase in number of people who use noisy products. And what about tomorrow?

THE NOISE LEVEL RISES

An obvious rise in the noise level is occurring. Some contend that a jump of 30 decibels, or one decibel per year, has taken place in the noise level of our cities in the past thirty years.[1] To date no major long term study has been conducted either to verify or refute this estimate. However, several isolated studies of transportation activity conclude that street noise has intensified. In Germany, for example, a survey team found that between 1938 and 1952 street noise in sections of Berlin and Düsseldorf rose 8 decibels.[2] Noise of this type approaches, and more and more frequently exceeds, maximum noise standards established for industrial work environments. Yet there is little concern to protect the public.

NOISE AFFECTS HEALTH

Noise can affect human health. The effects fall into two overlapping categories, consciously perceived and insidious effects. Consciously perceived effects are those recognized by the recipient or person. They usually have a subjectively irritating or a nuisance character, but they are nevertheless important functionally, producing such effects as interference with thought processes, communication disruption, performance impairment, sleep disturbance, and general mental stress.

Noise is also often an insidious pollutant. Subjectively we do not realize its deeper physiological impact. Hearing loss is a prime example. Initial sensory damage occurs in sensitivity to the upper frequencies, usually above 4,000 cycles per second. Although these frequencies are important to understanding speech, loss of awareness of them may not be noticeable to the victim. In fact, one may permanently lose up to 40 percent of hearing before noticing a loss which a medical examination will reveal. Besides causing hearing impairment, noise has been identified as a contributing factor in certain stress-related diseases, including

To what extent is external noise attenuated within a dwelling?

Is the population perceptually aware of local environmental problems?

To what degree and in what way is the population annoyed by noise?

What are the prevailing attitudes toward the question of noise affecting an individual's health and well-being?

Are past experiences important shapers of present attitudes?

Do any social variables influence attitude formation and responses to noise?

How useful are these variables in predicting a community response to noise?

Is human response to noise related to either the intensity of general community noise or specific noises?

Does the time, pattern, or frequency of noise occurrence affect the degree of community annoyance?

COMMUNITY
RESPONSE

In the final chapter I have attempted to construct a method for evaluating the health hazard of noise in a community and have described a model presenting the three approaches from which noise can be managed—the source, the path, and the receiver. Specific, practical suggestions for controlling aircraft noise—one of the greatest contributors to noise pollution—are offered, based on this model.

Acknowledgments

Many people have given me generous assistance throughout the writing of this book. I would like to thank the members of the Research Department at the West Philadelphia Community Mental Health Consortium who assisted in preparing and administering the social-environmental questionnaire, in particular Mary Bruce, Phillip Murray, Blair Kennedy, Larry Schein, and Dr. Lee Yudin. Dr. David Boyce, Professor of Regional Science, University of Pennsylvania, and members of the University Computer Center provided considerable help in the programing and analysis of the social data.

The instrumentation necessary for conducting the acoustical survey was kindly loaned by General Radio Company. Questions of a technical nature were capably handled by the staff of their Philadelphia area office and Karl Dolle, Department of Electrical Engineering, University of Pennsylvania. The City of Philadelphia and particularly Jesse Lieberman, Chief Occupational and Radiological Health Section, Department of Public Health, generously supported the study by providing members of his staff to assist with the acoustical survey. Mr. Lawrence Q. Rader, Chief of Airport Administration, Philadelphia, was also helpful in providing information. My sincere appreciation to Mr. Charles Foster, Director of the Office of Noise Abatement, U.S. Department of Transportation, and his staff for assisting me in my research. Mr. Richard H. Broun, Director of Environmental and Land Use Planning Division, U.S. Department of Housing and Urban Development and his staff were extremely generous with their time also. I would like to further acknowledge Mr. James Wright of the National Academy of Science who capably answered legal questions concerning noise and its control. Dr. Alvin F. Meyer, Director, Office of Noise Abatement and Control, U.S. Environmental Protection Agency, kindly provided information concerning the status of the Agency's noise program.

Members of my doctoral dissertation committee were most generous with their time in reviewing this manuscript. Mr. James Botsford, Noise Control Engineer, Bethlehem Steel Corporation, gave me professional advice within the subject area of acoustics. Without his capable and untiring help the scope of this study would have been limited. Dr. Larry M. Heideman, Professor of Community Medicine, University of Texas at Houston, made many thoughtful suggestions regarding research design and the medical implications of noise. I would also like to thank the dissertation supervisor, Dr. Stephen Ring, Eastern Pennsylvania Psychiatric Institute, for his guidance in the area of psychiatric epidemiology; and Dr. Morton Schussheim, Chairman, Graduate Group in City Planning, University of Pennsylvania, for his useful recommendations.

For her thorough and conscientious review of this manuscript and her moral support I would like to thank my wife Sarah.

Most sincere appreciation goes to the residents of West Philadelphia and Tinicum Township who so willingly gave their cooperation in the field of study.

March, 1971

Noise Pollution

The Unquiet Crisis

I

Community Noise as a Social Problem

The attitudes of society have allowed noise to become an environmental problem of sizable proportion, as later chapters will illustrate in detail. Society, through its technology, has created noise and has continued to tolerate it, with only sporadic attempts to control it. This chapter isolates nine social factors which contribute to our failure to meet the problem of noise pollution head on (see Figure 1.1.): the auditory regression of modern society, ignorance and underestimating the hazards of noise, our adaptation to noise, the priority of other urban concerns, institutional apathy (both public and private), conflicts in societal goals, the difficulty of establishing priorities in solutions to noise pollution, and unawareness.

In the development and growth of his sensory world, man has relied upon certain modes of perception as a means of comprehension. It seems that today visual perception is the predominant sense. Other senses (auditory, tactual, olfactory, and gustatory) are complementary but subordinate. The auditory sense, being less than fully used, has consequently become less sensitive. For example, visual cues often are more rapidly perceived than purely auditory cues. Popular expressions in use today reflect the present state of visual dependence. Such sayings as, "One picture is worth a thousand words," "Seeing is believing," "I am from Missouri, show me," and "I can't hear you without my glasses," illustrate this visual awareness.

Hearing is fundamental to the formal process of learning. In many aspects of education, visual learning with auditory reinforcement is the commonly accepted teaching method. Throughout antiquity and the

Auditory Regression of Modern Society

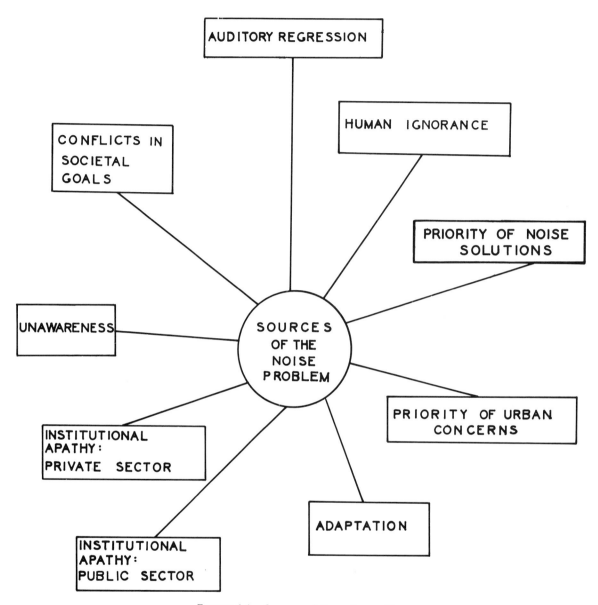

FIGURE 1.1 *Sources of the noise problem*

Middle Ages, reading was oriented to the ear. "Reading" meant "reading aloud." This method produced what Marshall McCluhan refers to as a synesthesia or interplay of the senses.[1] Throughout the Middle Ages, manuscripts were written with minimal punctuation and other visual aids. Word separation was largely ignored. This characteristic style was geared to the ear, not the eye, even as late as the seventeenth century.[2] Today we have moved away from using the auditory sense in reading with our emphasis on speed reading and visual scanning.

In general, learning became primarily visual with the advent of the scientific revolution. The sense of visual observation lent itself better to the scientific method. Touch, taste, smell, and hearing, described as affective senses because they are predominantly associated with the sources of emotional life, yielded to reasoning on grounds of visual observation and scientific experimentation.[3]

Man's entire way of life, not just the processes of reading and learning, is affected by how his environment is perceived. History suggests that major shifts in the sensory structure of perception have now occurred. In the Middle Ages, for example, man's auditory sense was keenly developed. In a world of little artificial light and nearly universal illiteracy, medieval man relied heavily upon his ears for information. He saw with his ears. Throughout that period, the church bell was not rung for esthetic purposes—it ordered men's lives and gave them meaning. The tolling of the bell evoked an encompassing range of emotional responses, from happiness and joy during holy days and festive occasions, to fear when the town was being attacked, and to sadness after battle when the death toll could be heard. The message conveyed was so precise that the bell's final strokes "indicated the age, sex, and social rank of the dead."[4] Another auditory vehicle important to the medieval townspeople was the crier or bellman. The town crier acted as a verbal newspaper, assisting those responsible for law and order in conveying their messages.

Because auditory communication was essential for community survival, steps were often taken to reduce noises that interfered with this vital function. By the thirteenth century, many towns had enacted laws prohibiting blacksmiths from working in the early morning hours because of bothersome noise.[5] In contrast with today, the population could enjoy relatively noise-free sleep from sunset to sunrise. Among those towns which had a thriving marketplace, paved streets were a major noise source. They became a particular nuisance when iron-rimmed carts entered the market towns from nearby farming areas. Laws were introduced to prohibit the use of these carts in the marketplace. In Beverly, England, a fine was imposed on persons driving iron-wheeled carts wherever stone pavement existed.[6] The less noisy and destructive wooden-wheeled carts could operate more freely in most towns. (In contrast, ground transportation noises are now legally controlled only in a few instances.) Although probably less intentional, the irregularly curved medieval street, by its design, reduced community noise propagation more than does the more recent rectilinear or grid street system.

After the so-called "dark ages" came the Renaissance and with it a new ordering of perceptual skills. Slowly the sense of sight replaced the sense of hearing as the principal means of perceiving. This trend has continued more or less up to the present.

How does this shift affect our lives today? In two ways: in our recognition of problems, and in their resolution. First, man's pre-occupation with what is visually recognized affects his awareness of environmental problems. Waste and litter, air pollution, and even water pollution are largely recognized as environmental maladies because they confront us visually. A darkened sky filled with particulate matter, a residential street cluttered with decaying abandoned vehicles, a vacant lot cluttered with garbage and refuse, a murky-looking stream containing solid wastes are not scenes to go unnoticed. Although no action may be taken to correct such conditions, there is at least visual recognition. Noise, unfortunately, does not assault our eyes, and consequently there is usually a tendency to tolerate it. However, when noise interferes with an activity which is primarily visual, community complaints are high. A sizable number of Tinicum residents were annoyed by aircraft because it interfered with television watching even though they could hear their TV. Still others complained about the aircraft noise problem in visual terms saying it soiled their home and property.

Recognizing the environmental problem is important. Equally important is finding the solution. Often a visual rather than an acousti-cal solution is used to solve a noise problem. Visual relief from a noise source can reduce the community complaint level. Utility companies are familiar with this phenomenon. When they erect fences or barriers around noisy substations, noise complaints usually lessen even though the noise reduction may be only a slight one. A similar response occurs around noisy roadways, where buffers consisting of fencing or land-scaping are installed. Here again, noise complaints seem to evaporate despite the lack of any significant sound reduction.[7] Visual screening is in keeping with our current way of problem solving, "If you don't see a problem, there isn't one."

Visually recognized environmental problems usually receive more prompt attention. Air pollution caused by aircraft has been given more rapid attention than noise pollution. States such as New Jersey, New York, Massachusetts, and Illinois are requiring airlines to modify or install smokeless engines. Those commercial aircraft using the Newark, New Jersey, airport must comply by October, 1971.[8] Jets will be smoke-less long before they are quiet.

The fault lies not only with the public, but also with the professions responsible for managing the environment. Consider the architect. Responsible for designing structures, he considers his task as principally visual. This "oversight" is largely due to a visual emphasis in architectural training. Although changes here are not to be expected soon, the architectural profession is beginning to be aware of the noise problem:

> If only a fraction of the effort applied to the visual aspects of a building were to be expended on acoustical considerations, the world would be a quieter place to live in, and at least one of the major causes of tension could be drastically reduced.[9]

The problem of auditory backwardness or regression is not universal. A segment of our population still uses sound for direction and consequently is very aware of noise as a pollutant. In some respects this group's predisposition to the sense of hearing is similar to medieval man's. The blind utilize their ears more fully, but this is by necessity. In their perceptual world the auditory sense is a vehicle to achieving personal well-being. For them the environment consists of various auditory patterns, some of which convey a feeling of warmth and security, others of which convey coolness or impersonalness.[10] For the rest of us the auditory sense has atrophied. The sighted population today seldom relies upon it even for their own protection and safety. The auditory comfort and enjoyment we could experience from the environment are now a missing vital dimension. Blind professionals, including architects, city planners and behavioralists, are needed to help society rediscover the auditory sense for their own environmental well-being.

Human Ignorance

Until the late sixties the prevailing attitude of urban society toward the environmental pollution problem was one of disregard, insensitivity, and misunderstanding. Interest has grown for the need to control air, water, and even land pollution. Comparatively speaking, however, noise has not yet gained the public's attention. The former Surgeon General speaking at the first Conference on Noise as a Public Health Hazard observed that attitudes toward noise were similar to those toward air pollution ten years earlier. "Back in 1958 people were saying, Air Pollution problem? I don't smell anything.' Today there are apologists for some of the noisier phenomena in our society saying, 'I don't hear anything.'"[11]

There are numerous examples of these attitudes toward noise.

NOISE AS POWER

People tend to equate the noise of a machine with its power. Without this accompanying noise, consumers often believe there is a comparable loss in power. A major manufacturer during the early 1960's introduced two identically powered lawnmowers. One had been acoustically engineered to operate more quietly than the other. There was a wide discrepancy in sales, with sales of the quiet one lagging decidedly, and it was later removed from the market. Why? According to the manufacturer, the buying public thought that it "lacked power" and did not operate as well as the louder, "more powerful" model.[12]

Another manufacturer is testing to see if this consumer attitude still prevails. The Whirlpool Corporation presently is offering the public two similar window air-conditioning units. Among other differences, one has been engineered to reduce the noise by means of a sound-absorbing decorator panel. The other is a standard model. Performance characteristics (cooling capability, operational costs, etc.) are comparable, though the purchase price is slightly higher for the quieter unit.

Often exhaust systems of vehicles are altered in the belief that loudness gives the vehicles substantially greater power. The trucking industry is probably the most flagrant violator, since many carriers, or truckers themselves, remove mufflers and modify the exhaust system. The problem lies more with the user than with the manufacturer, who generally provides for suppression of vehicle exhaust noise adequately.[13] The outcome is a much noisier operating truck having only a nominal increase in actual power. Whatever small gain there may be in engine power because of reduced back pressure (probably less than 3%) it is not proportionate to the increase in truck noise. The noise is a liability endangering the driver's hearing and performance, while also contributing to community din. (See Chapter III, pages 65–67.)

NOISE AS EFFICIENCY   Not only is there the attitude that noise means greater power, but also that noise contributes to greater operational efficiency. The vacuum cleaner is a good example. When it is being used the noise the housewife hears reinforces her belief that the vacuum is "performing properly." Conversely, a quiet one is perceived as inept, and therefore is not capable of cleaning as well. Midway through 1967 the Hoover Company introduced a vacuum cleaner decidedly quieter than the models of previous years. Its national advertising program described the appliance as "whisper quiet," powered by a one horsepower motor.* Housewives evidently were not convinced of the performance capability of the quieter model. They continued to purchase the noisier model, under the misconception that to clean properly a vacuum cleaner has to be noisy.[14] This notion may be reinforced by the fact that Consumers Union gives a top-rating to a cannister model with an ex-

---

*See Appendix B, page 199.

cessively high noise level.[15] A frequency noise analysis was performed after my wife complained of ringing in her ears, as well as an inability to hear telephone, baby, or front door while cleaning with it. The results confirmed the obvious. Under normal operation the noise reached 86 decibels (both dBA and dBC) while the speech interference level averaged 77 decibels (dBC), a level requiring shouting to be heard.* The marketing people at the Remington Division of Sperry Rand Corporation encountered attitudes similar to the Hoover Company when they introduced a quiet electric typewriter.[16] Noise control engineers had removed the "clacking" sound from the typewriter. However, secretaries complained that it was noticeably "slower" than the noisier, otherwise identical machine.

Consumer preference has been strong enough that, according to manufacturers of lawnmowers, vacuum cleaners, and refrigerators, noise at one time engineered out had to be rebuilt in. Companies that make refrigerators claim housewives want to hear the steady rumble of the motor.[17] This noise assures them it is running. Even some industries associate the sound level of equipment with efficiency. A "quiet" jackhammer has been on the market for several years, but "salesmen for competing products have succeeded in persuading contractors they are less effective and underpowered."[18] In truth, all of these quieted products were as efficient as their noisier counterparts.

Quietness, as these examples indicate, generally is foreign to the consumer. This country has undergone a history of technological conditioning. The early products available to the public were minimally engineered, especially regarding noise suppression. To buy most consumer items, such as power tools, household appliances, recreational equipment, and automobiles meant buying noise. This was an accepted practice for there were no alternatives. Today a choice exists. The accustomed noise of most products can be removed, or at least reduced, through engineering means.[19] (Usually reducing the noise increases the purchaser's cost, but as a rule no more than 5–10%.)

Some companies have taken the initiative by demonstrating to the public what is possible. Ingersoll-Rand has introduced a large portable air compressor which substantially controls a major source of construction noise. The "Whisperized" SPIRO-FLO model (DL9005) generates an 85 dBA sound level at a distance of three feet.† Compared with the seven conventional compressors now available it is at least 25 decibels lower.[20]

A consumer re-education program is also necessary. Montgomery

---

*See Chapter II, pages 79-80.
†See Appendix B, page 204.

Ward is attempting to change the noisy household image. Their entire line of Signature appliances is being marketed with the theme "Silence is golden." Reliability as well as noise-free operation is stressed:

> . . . you'll hear a beautiful nothing from our Signature appliances. No rattles. No whines. No malfunctioning moans. Because this year like every year we've built in the same quiet dependability. It's as essential to us as our appliance features could be to you.*

Consumers must assert their preference to buy quiet, if indeed that is their preference, especially in relation to other goals and services desired.

NOISE AND SOCIAL RECOGNITION

Probably the most popular use of noise is to gain recognition and confidence. Unfortunately, many people attach a personal value to noise. It is an attention-seeking device that at least temporarily gives recognition and identity to its user. The greatest users today are youth. Unmuffled cars or motorcycles signal their arrival and departure to and from the scene. Such cacophony gives them a feeling of being part of the "in" or "hip" crowd. It can also be interpreted as a protest against the "establishment's" highly organized, dull, quiet world. Motorcycle noise has been identified as a factor contributing to an emotional ailment: the motorcycle syndrome. Cyclists suffering from this apparent psychiatric disorder consider the noise emotionally gratifying. As one patient has stated,

> the noise is all you hear . . . there is a strength and power in it—52 horsepower. It's masculine and makes me feel strong. I approach a girl on a cycle and I feel confident. Things open up and I am much more at ease.[21]

Loudness, often outright noise, is also linked to the music of the younger generation. The world of electronic amplification has found a place with this age group. Loud music is considered by youth to be an identifiable trademark. Unfortunately, they do not give equal consideration to the very real problem of hearing loss. Temporary, as well as permanent, hearing loss is being found by audiologists who have examined musicians who frequently play electronically amplified music.[22]

Although it seems our tolerance for noise is inexcusable behavior, there is a good explanation. We are culturally conditioned. Noise is an integral part of our "way of life." It is not approached critically, as an enemy or foreigner during the formative years of growing up. As parents we introduce our children to toys that are needlessly noisy; certain ones are even hazardous to hearing.

*See Appendix B, page 199.

Some toy motorcycles, trucks, and tractors have a built-in roar. There are push-type lawnmowers that purposely make a buzzing sound to imitate their adult counterparts.[23] Then there are pump guns, cap pistols and the firecrackers that generate an impulsive type noise hazardous to a child's unprotected ear.[24] We allow children to play with some cap pistols having a higher noise level than a 45 caliber pistol, 22 caliber rifle, or an M-16 rifle utilized in combat. Not only do parents buy such toys, but noise and its possible hazard is seldom mentioned. How often do we point out examples of man's impressive technology to our offspring without mentioning their faults? We permit, even encourage, our children to marvel at the airplane, but do we ever mention the noise and air pollution problem?

Hopefully this situation will improve and parental awareness will improve and parental awareness will develop. The earliest books for children must begin telling how the environment really is, what is wrong with it, and what is necessary for improvement. *The Listening Walk* is a type of book headed in this direction because it describes the environment in which the youngster lives, including the jet:[25]

> Sometimes my father and I take Major to the park
> I like it in the park
> It is cool and shady
> It is quiet there until a jet flies over
> Jets are very noisy. A jet goes
> eeeeeeeyowwwoooooooooooooo.

and the lawnmower:

> I listen to lawnmowers
> Lawnmowers are noisy.
>
> A power mower has a motor
> The motor makes a steady zooming noise
> It goes like this:
> z-z-z-z-zzzzzzooooooooooooooooooommmm.

Some parental interest is starting to appear. Citizens for Clean Air, a New York based group, has begun protesting toys that imitate actual air pollution conditions. A local department store selling a play truck called "True-Smoke" was picketed by twenty-five CCA members carrying placards stating "Keep Pollution Out of the Playroom." In an interview, the organization's executive director commented, "We are not so much objecting to a toy as we are to an idea—in this case the notion propagated by the manufacture and sale of such a toy, that pollution is an acceptable way of life."[26]

*Our Adaptation
to Noise*

The adaptation of an organism to environmental conditions is a common occurrence throughout history. But adaptation is a two-edged sword. It can often be a saving grace, but at the same time it may have created a certain false sense of well-being.

On the positive side, man's ability to adapt has allowed him to survive under conditions in which, without this ability, the odds of his survival were slim. The early westward migrants adapted to changes in climate, diet, and life style in order to meet the demands of settling new areas of the country. More recently, the urban population has adapted to a series of environmental irritants (air, water, refuse, and noise pollution) in order to endure city living. These poor conditions have often been mentally suppressed or disregarded by the population in order to function normally.

On the negative side, our ability to "tolerate" certain environmental predicaments has lured us into a false sense of well-being. Surely the air is not any cleaner simply because we don't think about air pollution in our major cities. We still have to breathe it, and the fact remains that it adversely affects human health. Unfortunately for our physical health, the adaptive technique of mental suppression has no physiological parallel.[27]

The same phenomenon is also valid for noise pollution. Subjective responses to noise are not always an accurate indication of its health effects. Generally, about one fourth of the urban population is never annoyed by noise regardless of its intensity, according to one acoustics consultant.[28] A similar finding has been reported in studying human response patterns to aircraft noise.[29] Even at the noisiest monitoring stations in our field survey, where aircraft predictably disrupted sleep during the night, some residents commented that they had become "accustomed" to the noise.

Physically there is little adaptation by the human body to noise, especially to intense noise. Most adaptation is mental. Although noise may be damaging to one's physical health, the individual makes the necessary mental adjustment. This type of adjustment is common when sleep is desired, even though physiologically the quality, duration, and recuperative value of sleep are affected by the noise (see Chapter 3).

The obvious extension of this current adaptive trait is the loss of our ability to recognize the potential hazards of noise, so that we no longer react to it. This could lead to even more detrimental effects on our health than we are now experiencing. It is conceivable, particularly among the younger generation and its progeny, that we eventually will adapt automatically to the prevailing environment. We can ill afford to let this happen with noise. The present amount of hearing loss discovered among adolescents is a serious warning for the future.[30] Larger and larger portions of the world are infested with noise, and there are fewer and fewer quiet places. Comparisons between noisy and non-noisy environments are becoming more difficult, particularly within developed nations.

Fortunately, cultural differences still prevail. The countries of Europe—especially Germany, France, Britain, Sweden, and Switzerland—are less tolerant of noise. A natural resource delegation from the Department of Interior visited Germany and found the German people intensely concerned about protecting their environment.[31] Strongly supported by its citizens, Germany has taken major steps to abate noise. It has adopted with considerable success the most stringent noise regulations in existence.

American businesses having overseas operations find this to be true. Their environmental control expenditures abroad, and especially within Europe, are larger than in the United States. Even military operations are monitored carefully with respect to noise. Military vehicles must meet more severe noise emission criteria abroad than at home. In one instance an entire shipment of U.S. Army ambulances manufactured in Detroit had to be modified because they exceeded French noise requirements. Noise associated with at least one Army training installation in Germany exceeded a community noise law.[32] The author, as part of an environmental survey team, recommended altering the use of the facility since a local community noise annoyance problem did exist. The Army has not had to comply with a law having such stringent requirements at home. Clearly there are various degrees of environmental awareness and of adaptation to environmental problems among various cultures. The chances of noise becoming a major environmental problem are much smaller in Europe than in the United States.

Noise represents just one of many problems besieging urban society. It has been overshadowed by burgeoning crime rates, racial disharmony, student unrest, civil strikes, and the Southeast Asia conflict. These constitute major problems that are demanding attention for the allocation of federal resources. Education, social welfare, medical assistance, and housing are some of the needy areas receiving financial support. Expenditures for these purposes, combined with the more fixed ones (national defense, veterans benefits and services, income security, and debt retirement account for 76 percent or $150 billion of a $198 billion federal budget) restrict funds available for other purposes.[33] Consequently there is little economic support at any governmental level for tackling existing environmental problems (see Table 1.1). In fiscal year 1970 the federal government obligated only $165 million for air pollution control. This was to be increased slightly in 1971

Priority of Other Urban Problems

TABLE 1.1.   U.S. Federal Budget, Fiscal Year 1971*
                         (in billions)

Fiscal year 1970	Fiscal year 1971	Budgetary item
$197.885	$200.771	Total budget
149.766	150.241	National defense, Veterans benefits, Interest, Income Security
.828	1.636	Water pollution control
.165	.195	Air pollution control
.034	.040	Noise pollution control

*Preliminary data, August 6, 1970. Executive Office of the President, Bureau of the Budget, Office of Management and Budget.

to $195 million. In water pollution control the 1970 federal allocation was $828 million, and this figure is nearly doubled for 1971. Federal support for noise abatement is much smaller, amounting to $34 million in 1970. On a comparative basis noise control receives one fifth of what air pollution receives, and one fortieth of what water pollution receives in fiscal year 1971.*

The majority of noise control funds are being spent on aircraft, which represents just one of many community noise sources. In fiscal year 1970, $31 of the $34 million budget was devoted to subsonic and supersonic noise.

Another complication in obtaining adequate funds for controlling noise is the fact that there has been no federal noise legislation comparable to the Clean Air Act or the Water Pollution Control Act expressly authorizing appropriations. However, the Senate Clean Air Bill (S-4358), sponsored by Senator Edmund Muskie, contained a section authorizing $30 million for establishing and operating an Office of Noise Pollution Control in the Department of Health, Education and Welfare. It was slightly modified because the House bill (HR 17255) became public law. Title IV of Public Law 91-604, cited as the "Clean Air Amendments of 1970," contains provisions for establishing an Office of Noise Abatement and Control.[34] Approved by Congress on December 31, 1970, this office is authorized to appropriate an amount not to exceed $30 million. However, to date no funds have been appropriated, but federal approval is expected by early 1972.

*The proposed budget in 1972 for noise pollution control is $66 million (Office of Management and Budget, U.S. Bureau of the Budget, "Federal Environmental Programs," Washington, D.C.: GPO, 1971, p. 222).

Only by executive order did former President Lyndon B. Johnson request federal departments and agencies to begin participating in a unified effort to try to solve just one part of the problem, aircraft noise.[35] In 1967 he established the Federal Interagency Aircraft Noise Abatement program. But the growth and expansion of this program is not assured under the Nixon administration. Relative, then, to the total package of environmental problems, noise has been given a rather low priority.

The federal government is spending as much to compensate those affected by noise as it is spending to control noise. Although current statistics are not available, in 1967, at the national level, over $35 million was awarded in noise-related compensation claims (see Table 1.2). The largest yearly expenditure is for compensation claims for loss of hearing. Our over 90,000 veterans with service-connected hearing disability received approximately $32 million in 1967.[36] In addition, federal civilian employees were awarded $1 million because of noise-induced hearing loss. Besides outright compensation, medical treatment is available, including medical and audiological evaluations, hearing aids, batteries, instrument maintenance, and transportation to and from medical facilities. Authorized under the VA Compensation Act, these benefits in 1967 cost the U.S. Government another $1.4 million, excluding VA employee salaries. The settlement of sonic boom claims associated with military supersonic flights from 1956–1967 amounted to $1.3 million; $145,000 for 1967.[37] This federal total could be considerably larger than just $35 million because of the following:

(a) not all federal employees eligible for compensation, due to noise-induced hearing loss, file claims
(b) not all people who incur damage caused by supersonic overflights file claims
(c) the budget available for settling sonic boom claims is fixed; consequently, when the budget is exhausted, claims are no longer processed (in contrast, the compensation system in Great Britain allows each case to be judged on its merits and there is no fixed budget)

Cost of Noise: Federal Level (1967)   TABLE 1.2.

Hearing compensation	
Veterans' compensation	$32,680,000
Civilians' compensation	1,000,000
Medical treatment	
Veterans	1,400,000
Sonic boom compensation	145,000
TOTAL	$35,225,000

(d) the millions of dollars awarded to community residents near airports could be considered a federal cost since the federal government is responsible for controlling aircraft and its noise. Fortunately for the government, usually the airport operator or airline is held liable.

There are, however, indications that among the population noise pollution is considered to have a higher priority than government has assigned it. In other words, a discrepancy exists between the government's activity and the citizens' concern. Our observation is based upon several community opinion surveys in which noise was compared to other community conditions. The London survey of 1961–1962, conducted in the vicinity of Heathrow Airport, found that noise ranked high on the list of "The one thing people most wanted to change."[38] A more detailed, comparative analysis of social and physical environmental conditions in West Philadelphia brought similar findings,[39] in addition to our field data presented in Chapter 5.

Election results across the country in 1970 definitely showed that the public is displeased with current budgetary priorities. The allocation of funds for environmental protection should experience a healthy growth. Political candidates who are conservation-minded drew strong public support during these elections. At the national level it appears that

> Support or opposition from new environmental activist groups appeared significant if not pivotal in a score of contests for the House of Representatives and at least one Senate race.[40]

Hopefully, with the growing environmental awareness among elected officials, as priorities do change noise and the need for its control will receive greater financial recognition. The enactment of the 1970 Noise Pollution and Abatement Act, as Title IV of Public Law 91-604, could be a major stimulus for changing present priorities.

Institutional Apathy: Public Sector

The prevailing institutional structure of society (political, educational, legal, corporate, etc.) has responded with indifference to the noise pollution challenge, a challenge directly affecting the quality of urban life. Both the private and public sectors of society are responsible for institutional inactivity. Certainly, no adequate thrust and direction are supplied by any governmental echelon, though the government is the institutional representative within the public sector. The more than 80,000 units administering federal, state, and local government in the United States[41] are collectively failing to offer solutions. This inactivity is caused by their organization, by legislation, and by implementation problems.

Most structures of government are organizationally incapable of confronting problems of an interdepartmental nature because such problems do not fit into the traditional table of organization. Effective solutions require either close interdepartmental cooperation under the present setup, or alternatively a revision of governmental organizations. The latter alternative means widening the scope of a city department or agency so that it will be able to look at a total urban problem rather than a mere segment of one. Many different disciplines would then be grouped under one "roof;" the resulting interchange and "cross-fertilization" might offer fresh insights and generate imaginative solutions.

A change in organizational structure has taken place in New York City. Mayor John Lindsay in 1967 created what some call a superagency to handle in a comprehensive manner the massive environmental quality problems facing the city.[42] Officially titled the Environmental Protection Administration, this superagency is designed to monitor noise pollution as well as other environmental health matters. Noise problems are being handled through the agency's Bureau of Noise Abatement. Although beginning slowly, the Bureau has drafted a city-wide noise control ordinance.[43] This agency represents a significant shift from traditional organizational design.

In most cities, antiquated governmental machinery prevails. Within local departments of public health, noise—if it is considered at all—is defined solely as an industrial worker problem rather than as also a community problem. Public Health officials assigned to the noise problem are usually drawn from an industrial hygienist's background. Their training deals almost exclusively with industrial work environment; consequently, they are not likely to give much consideration to the community environment outside the factory.

There are other organizational handicaps. The responsibility for noise control is frequently a "paper function" of some department or agency. Such is the case in Philadelphia, where this responsibility was shifted from the City Department of Public Health to the Police Department. When Public Health was in charge of community noise, several of its members generously contributed of their time and energy to the Mayor's Committee for Noise Abatement. The subsequent transfer of this function to the Police Department led to the dissolution of the Committee. Equally important is the fact that community noise control is now given token attention by the already overburdened police. Usually "action" is taken only when a complaint is received, meaning that the alleged noise complaint is acknowledged and entered into a log book. Community noise research is no longer undertaken, and the

GOVERNMENT
ORGANIZATION

earlier noise abatement campaigns informing the public about this menace have been dropped. Philadelphia is not alone in pigeonholing noise control. There are other examples of governmental apathy, among them St. Louis, Missouri, where the city also has a legally stated responsibility but is doing nothing.[44]

President Nixon recommended a reorganization of the federal agencies responsible for controlling the environment.[45] Most pollution control activities now have been transferred to an independent Environmental Protection Agency (EPA). The consolidation provides the means for a more concerted effort to improve the environment. Five categorical offices have been established by the EPA Director, William Ruckelshaus. They are Water Quality, Air Pollution Control, Solid Waste, Radiation, and Pesticide.

The Office of Noise Abatement and Control is located within EPA. This office has a director but few staff and rather than being an independent office it is part of the Planning and Management Office. Because of various departmental interests, the federal noise program may remain divided among several departments, rather than being more effectively consolidated.

INADEQUATE
LEGISLATION

Contributing to organizational ineptness is the failure of policy makers to initiate action leading to legislation. Local governing bodies generally have not legislated against noise pollution. When action is taken, city councils approve ordinances prohibiting the occurrence of community noise as a nuisance. Probably over 500 such ordinances can be found throughout our country. Their effectiveness, however, is limited for a variety of reasons. A large number of ordinances define noise nuisances so narrowly that such legislation applies to only a small part of the problem (i.e., loud or noisy mufflers). Either by design or because of inadequate technical and legal advice, city councils frequently prepare ordinances of little legal value. Ordinances of this type are written so vaguely that they cannot be defended. Rather than prescribing a maximum allowable decibel level for a noise source, a law may read, "No person shall operate any vehicle which causes unnecessary noise." The legal question, of course, is what constitutes "unnecessary noise." Frequently this is not quantitatively defined. In other situations, a maximum decibel level is indicated, but it may be so low that normal talking on the street is a violation,[46] or so high that all noise sources in a city would be legally immune. A final point is that some local governments have approved noise ordinances dealing with areas in which they have no legal jurisdiction. According to a federal court opinion in a Hempstead, Long Island, case, no municipality has the legal authority to control the noise of aircraft overflights.[47] Although this decision is still being appealed, it seems that the management of navigational airspace (which includes noise control) is a federal responsibility.

Besides nuisance-type legislation, some municipal governments have adopted, or are in the process of adopting, building codes including sections dealing with noise transmission. Building codes have been more widely neglected than community noise regulations: before 1968 no major city had ever passed a building code requiring acoustical materials that limit noise transmission within a building. In late 1968, New York City became the first major city in the United States to approve a building code with such a provision.[48] Outside of the United States, particularly in Scandinavia and elsewhere in Europe, there are very sophisticated and stringent building codes that minimize the noise transmission problem. Many of these foreign regulations date back to as early as 1938. Building regulations in England and Wales are very thorough and include defined grades of sound insulation both for walls and for floors between dwellings.[49] Both grades surpass those presently in use by the building industry in the United States.

Municipal governments are not the only ones who have lagged in noise legislation. State governments have been equally negligent. In fact, their major contributions have been workmen's compensation laws (i.e., determining noise-induced hearing-loss criteria applicable *after hearing loss has occurred*).

The states of California and New York are two major exceptions. In 1967, the California legislature adopted two Vehicle Code Sections, 23130 and 27160, which establish maximum permissible noise limits for vehicles.[50] The latter section applies to new vehicles sold in the state since January 1, 1968. It states:

No person shall sell or offer for sale a new motor vehicle which produces a maximum noise exceeding the following noise limit at a distance of 50 feet from the centerline of travel under test procedures established by the department:

(1) Any motorcycle manufactured prior
     to 1970   92 dBA
     after 1969 and before Jan. 1, 1973   88 dBA
     on or after Jan. 1, 1973   86 dBA

(2) Any motor vehicle with a gross
     vehicle weight rating of 6,000
     pounds or more manufactured
     after Jan. 1, 1968 and before
     Jan. 1, 1973   88 dBA
     on or after Jan. 1, 1973   86 dBA

(3) Any other motor vehicle manufactured after Jan. 1, 1968 and before Jan. 1, 1973	86 dBA
on or after Jan. 1, 1973	84 dBA

Section 23130 of the California Vehicle Code establishes maximum permissible noise limits for vehicles operating on highways.

No person shall operate either a motor vehicle or combination of vehicles of a type subject to registration at any time or under any condition of grade, load, acceleration or deceleration in such a manner as to exceed the following noise limit for the category of motor vehicle based on a distance of 50 feet from the center of the line of travel within the speed limits specified in this section:

	Speed limit of 35 mph or less	Speed limit of more than 35 mph
(1) Any motor vehicle with a manufacturer's gross weight rating of 6,000 pounds or more after Jan. 1, 1968 and before Jan. 1, 1973	88 dBA	92 dBA
on or after Jan. 1, 1973	86 dBA	90 dBA
(2) Any other motor vehicle and any combination of vehicles towed by such motor vehicle after Jan. 1, 1968	82 dBA	86 dBA[51]

California is also working on other noise fronts. The State Aeronautics Board has approved and sent to the Legislature a 15-year airport noise abatement program. The regulation applicable to all civilian airports stipulates the average airport noise level from 1972–1975 would be 80 decibels. By 1986 this would be lowered to a Community Noise Equivalent Level of 65 decibels. Noise monitoring equipment would be required at these airports and the aircraft violating this noise limit would be subject to a fine not to exceed $1,000.[52]

Under the 1970 Environmental Conservation Act New York State established a Department of Environmental Conservation. Provisions of the act grant the authority for a statewide noise abatement program.[53] In Illinois a Bureau of Noise Pollution Control has been established in the State Environmental Protection Agency.[54]

Other states, including Connecticut, Maryland, New Jersey, North Dakota, and Pennsylvania have introduced bills into their respective legislatures to broaden government responsibility in noise pollution control.

The federal government, like local and state governments, has largely ignored the problem of urban noise. Congress has approved just one public law dealing with the subject.* This law authorizes the Federal Aviation Agency to prescribe and amend rules and regulations for the control of aircraft noise and the sonic boom.[55] Although very important, this law is concerned with only one segment of the noise problem. Furthermore, the FAA has been responsible for managing navigable airspace since 1958, but it was all of ten years before any agency received legal authority to control aircraft noise. Up until 1968 all U.S. aircraft had federal permission to make as much noise as they cared. The nation's first federal regulation limiting the noise of new commercial and civil aircraft was promulgated by the FAA in late 1969.[56] By amending federal aviation regulations, the FAA prescribes noise standards for aircraft type certification. The maximum noise levels for subsonic transport and turbojet-powered airplanes are measured at three stages of flight:

1. during takeoff, 3.5 nautical miles from the start of the takeoff roll on the extended centerline of the runway;

2. on approach, at a point 1 nautical mile from the threshhold on the extended centerline;

3. after liftoff, at the point, on a line parallel to and 0.25 nautical miles from the extended centerline of the runway, where the noise is greatest, except that, for airplanes powered by more than three turbojet engines, this distance must be .35 nautical miles.[57]

Depending upon aircraft weight, number of engines, and point of measurement, the maximum allowable noise level ranges from 108 EPNdB (Effective Perceived Noise Level in Decibels) to 93 EPNdB. (EPNdB is defined as the value of PNdB adjusted for both the presence of discrete frequencies and the time history.)[58] All new subsonic jet-powered aircraft must stay within these limits or they will not receive the government flight certificate necessary before entering service. Major loopholes do remain though. Present aircraft, including the noisiest plane in general use (Boeing 707) are exempt from the law.[59] (Plans are underway to extend its working life by installing a new wing assembly and consequently allowing the high noise level to remain unabated for an extended time period).[60] Second, the 747 will not be required to comply to the noise certification law until December, 1971. By then 90 percent of all these planes will already have been built.[61] Although noise certification of jet aircraft constructed since 1970 is a positive step, the noise levels airline manufacturers are required to meet certainly do not remove the community noise annoyance problem.

Furthermore, there are no FAA controls limiting the noise of rotary

---

*Congress is expected to approve the noise control act drafted by EPA by early 1972.

wing (i.e., helicopters), Vertical Take-Off Landing (VTOL), or Short Take-Off Landing (STOL) aircraft and this group is gaining popularity in our already noise-infested cities. At least one city, Berkeley, California, is restricting the use of helicopters because of their noise.[62] Plans for retrofitting engines on existing aircraft with some noise suppression control is only now being talked about by the FAA. They recently sought comments on a proposed rule establishing noise retrofit requirements for civil airplanes.[63] But future improvements do not seem promising. When queried about retrofitting our present jet fleet and establishing noise limits for helicopters, STOL, and VTOL craft, a federal noise abatement official commented that the aerospace industry could not afford other noise restrictions. If the federal government is not interested in regulating this industry, and by law no other governmental unit can, then no one is protecting the public's interest. A coalition of the airline industries with support from certain FAA officials and the Chairman of the Civil Aeronautics Board may well succeed in blocking any meaningful noise abatement program for existing aircraft.[64]

INEFFECTIVE IMPLEMENTATION

Implementation is the final ingredient necessary for the adequate control of noise. Without proper manpower and equipment, for example, the best-intentioned law remains ineffective. England has amended her motor vehicle code establishing maximum permissible noise levels. To enforce this amended code requires that the police monitor vehicular movements with sound level meters. A violator can be fined as much as £50 (approximately $120) for his first offense; the penalties, then, are relatively severe.[65] However, in the first three months of operation, no one has been prosecuted. Less than 10 percent of the police forces responsible for enforcement have sound level meters. The chief constables, already overburdened with other duties, feel that an inordinate amount of time is required for something one person has called "a load of rubbish" which is totally unenforceable.[66] Many constables merely wait and see, to learn how the police forces using the sound level meters will fare, and what problems they may encounter. For the time being, the Ministry of Transportation's new regulation is only a paper law.

A somewhat similar situation has developed in New York State. There, the state police are responsible for enforcing an anti-noise law which establishes maximum vehicular noise emissions permissible while traveling on the New York State Thruway.[67] The Thruway Noise Abatement Committee, organized in 1961, fought five years for its passage. Approved in July of 1965, the law was heralded widely. Speaking for proponents of the bill, one senator stated: "We have, at last, a means of enforcing our anti-noise laws against the small minority of offenders who have persistently ruined the sleep of hundreds of Westchester residents."[68] In retrospect, these platitudes seem to have been spoken prematurely. Little enforcement has taken place. Among trucks alone the average traffic volume on the Thruway is about 1,000 vehicles per hour, or 8.7 million trucks annually. However, the number of summonses issued for noise violations in two years has been six.[69] A major obstacle, as in England, is the lack of sound level meters. There are just two for the entire 400-plus miles of New York State Thruway. The police authorities apparently consider the enforcement of this law a burden; in any event, they have not been anxious to request further equipment. In both instances, it is doubtful that the public enjoys more protection with these laws than without them.

Not all motor vehicle noise ordinances are quite so poorly enforced. The Memphis, Tennessee, Anti-Noise ordinance, adopted in 1938, is a well-known exception. In this community, the Police Department takes great pride in its manner of dealing with offensive vehicles. During 1966, the Memphis Police Department made 5,760 arrests for the operation of vehicles with excessively noisy mufflers, and 360 arrests for other related violations.[70] Noise is defined as that which exceeds 90 dBA.

California is beginning to enforce the noise provision of its motor vehicle code on a statewide scale. The California Department of Motor Vehicles has six sets of sound measuring equipment, used by six teams of trained enforcement officers.[71] The results for the six-month period ending September 1969 were very encouraging (see Table 1.3). According to more recent discussions with Department of Motor Vehicles personnel, the enforcement program remains very active.

Problems in implementation are not limited to motor vehicle codes, and the police should not bear the full brunt of our criticism. Equal

California Motor Vehicle Violations   TABLE 1.3.
due to Noise: April—September 1969

	California Noise Enforcement Summary Report		
Type of Vehicle	Number checked	Number violations	Percent in violation
Under 6,000 lbs. GVW*	224,462	244	0.1%
Over 6,000 lbs. GVW*	112,958	1,089	1.0%
Motorcycles	1,693	34	2.0%
TOTAL	339,113	1,367	0.4%

*Gross vehicular weight.

fault lies in many other places. City planning bodies are notoriously insensitive to the need for controlling noise.

> The major obstacle to effective planning of roads and communities, whether for the sake of minimizing noise nuisance, for easing traffic congestion or for any other public advantage seems to lie in the fact that the authority of planning bodies is permissive in nature.[72]

Permissiveness on the part of the New York City Planning Commission in 1967 paved the way for the construction of two high-rise apartment towers directly under the approach to LaGuardia Airport's all-weather instrument runway. These towers, now completed, house 816 persons who are exposed to aircraft noise "equivalent to a diesel freight train traveling at 50 miles an hour passing at a distance of 100 feet every 45 seconds."[73] The decision of the New York City Planning Commission is of course contrary to the federal government's desire to create more compatible land uses (principally nonresidential uses) around airports in order to provide a needed buffer against noise. To assist in implementing this objective, a Technical Advisory Committee to the City Planning Commission was established in 1963; however, this committee had not met once in the two years prior to that regrettable decision.[74]

New York City's Mayor Lindsay asked the Planning Commission to consider the feasibility of heliport sites in Manhattan, close to a large population. There was formal opposition to the plan. Dr. Detler W. Bronk, President of Rockefeller University, had serious reservations. He felt that the noise could well result in injury to the school's hospital patients, and in a reduction of the accuracy of its research.[75] Originally, the land in question was to be used as a park and recreational area. Ultimately the city, with the commission's recommendation, granted Pan American airlines the permission to operate the heliport at 61st Street. The decision was hailed by the mayor as "an important step toward relieving air-traffic congestion at the major airports by making the smaller fields more attractive, particularly for corporate aircraft."[76] Neither the mayor nor the Planning Commission commented on the noise resulting from this operation, consequently helicopter service is still being provided. Serious consideration is also being given to the introduction of short take-off landing aircraft into the New York City area.

To suppress aircraft noise experienced by homeowners living near London's Heathrow Airport, the British government approved matching grants to insulate their homes. Evidently this program did not prove effective. A year after enactment, only 2,589 of the 60,000 householders eligible had inquired about the noise insulation schemes.[77] In this same period, less than half of these grants had been paid out. This experience illustrates another problem of implementation. The amount offered is not a large enough incentive. Most homeowners feel that these grants available through the British Airports Authority are not adequate, because they must pay at least one-third the cost of the job; this usually does amount to more than $300.[78] Many believe that, not being directly responsible for the noise, they should not have to pay to abate it. (A similar viewpoint was found among the Tinicum population; see Chapter VI.) If such grants are to be effective, it appears that a greater share of the cost will have to be borne by the government or the noise generators (airlines, manufacturers, and airport operators).

Implementation problems are not limited to ground and air transportation. In St. Louis County, Missouri, the zoning ordinance which stipulates residential noise levels is not enforced because the county does not own the proper sound measuring equipment.[79] Similarly, in New York City the police who can issue summonses for excessive construction noise rarely do so since they have no sound measuring equipment.[80]

We are failing not only to quiet the adult world, but also the children's world. Congress passed the Toy Safety and Child Protection Act, after overcoming a strong lobby, in November, 1969.[81] However, it was over a year before the Food and Drug Administration implemented the act. One week before Christmas (December 19, 1970) the FDA finally took its first step by banning, among other things, the sale of:

> Caps (paper or plastic) intended for use with toy guns and toy guns not intended for use with caps if such caps when so used or such toy guns produce impulse-type sound at a peak pressure level at or above 138 decibels, referred to 0.0002 dyne per square centimeter, when measured in an anechoic chamber at a distance of 25 centimeters (or the distance at which the sound source ordinarily would be from the ear of the child using it if such distance is less than 25 centimeters) in any direction from the source of the sound.[82]

It is not clear how the banning of noise-hazardous toy guns and caps will be implemented. These supposedly "banned toys" are still available, to the author's best knowledge.

Private institutions in our private enterprise system seem at least equally insensitive to the need for controlling noise. Spokesmen for industry frankly admit that changes will come about only with public insistence for consumer improvements. The primary objective of private enterprise is to maximize profit. Any deterrent to that objective, including the use of additional research-and-development funds for noise suppression, industry contends, raises the unit cost and places the enterprise at a competitive disadvantage. Although there are few

Institutional Apathy: Private Sector

examples, businesses have changed their tune when government, under pressure from its constituents and public interest groups, has demanded noise-abated municipal equipment.

In ordering additional garbage trucks, New York City wrote into the specifications that these trucks could not exceed a prescribed noise level in operation. Not all truck manufacturers shied away from this specification, even though the industry at that time had no model in production that was capable of meeting the requirements. At considerable expense to themselves, General Motors provided a so-called "noise package" for this order, and became the successful bidder.[83] Not only was the maximum passby noise level for the treated vehicle (measured at 50 feet) reduced from 88 to 77.5 dBA,[84] but the additional cost per unit amounted to only $102. Compared to the truck's total cost ($13,000), the added expense was nominal—less than 1 percent. By July 1972, New York City will have replaced their entire noisy fleet with quieter refuse collection vehicles.[85]

The trend toward specifying quieted or noise-treated items is also found among industries themselves. Some companies follow a "buy quiet" policy wherever feasible. Bethlehem Steel Corporation, when purchasing large-scale mining equipment, for instance, stipulates certain noise requirements.[86] Their purchase orders specify maximum operator noise levels equivalent to 90 dBA.

> The octave band sound pressure levels at the operator shall not, under any conditions of operation, exceed 85 decibels in any octave band above 300 cycles per second. Inspection prior to shipment from the factory shall be scheduled at the option of the purchaser to determine whether these noise requirements are met.[87]

Standard equipment such as a heavy duty bulldozer can meet these requirements if exhaust system and cab interior are modified. The employer directly benefits by controlling noise. There are fewer workmen's compensation claims because the equipment no longer is hazardous to hearing. Employee turnover becomes less of a problem, since the work environment is not as noisy. Indirectly the community benefits. When the equipment is quieted at the operator's site, less intense noise reaches the neighboring community.

A few corporations feel they have a public responsibility to be a good neighbor and that community noise control is part of the cost of doing business. Often a policy such as this can result in sizable savings to the corporation. At Esso the problem of noise is considered when a new refinery is being planned, rather than after it is in operation. This way there are fewer community complaints and lawsuits that require additional personnel and expensive litigation. The corporate image remains positive because there are good community relations. Furthermore, the cost of incorporating acoustical controls into the original design and construction typically amounts to one-half to 1 percent of the refinery's total cost. However, after a building is erected acoustical alterations are a major expense. If noise control measures are incorporated after the refinery is in use, costs may increase to approximately 10 percent, or one-tenth the building's original cost.[88]

Unfortunately these examples are exceptions to current private enterprise practices. The airport noise problem is one of many areas where good neighborliness seems to have been compromised if not ignored. This existing situation, as stated in the congressional investigation of aircraft noise,

AIRCRAFT INDUSTRY

> . . . can be characterized as one of conflict between two groups—those who benefit from air transportation services and people who (individually and collectively) live and work in communities near airports. The conflict exists because social and economic costs resulting from aircraft noise are imposed upon certain land users in the vicinity of airports who receive no direct benefits. It is important that this situation be rectified in an equitable manner consistent with the public welfare and the orderly development of air commerce.[89]

How has this situation been rectified by the aircraft industry? Before jet aircraft were introduced commercially, the industry spent roughly $50 million on research-and-development to perfect an in-flight sound suppressor for jet engines. By 1965, the investment had reportedly risen to $150 million.[90] Since the introduction of the turbofan engine, "suppressor-equipped jet engines were replaced" at a conversion cost of about $1 million per plane.[91] Despite this impressive inventory of costs, only moderate noise suppression has been achieved. Noise remains a major aircraft problem, and the public is the victim of certain questionable practices.

Certain airline operator practices indicate that the public is being victimized. One such practice is the "beat the box" game which, said *The New York Times* in 1967, is being played by most commercial airline companies using the airports operated by the New York Port Authority.[92] The Authority has established maximum allowable noise levels for aircraft departures at its installations. Jet takeoffs are not allowed to exceed 112 PNdB. To insure that limit is observed, ten permanent noise monitoring stations (called measuring boxes) are located at JFK, LaGuardia, and Newark Airports. Whenever a plane exceeds 103 PNdB this is recorded in the Noise Room operated by the Authority. However, the airlines have devised a way to "beat the box" so that they do not violate the 112 PNdB ceiling. Pan American Airways offers it as a contractual service to other airlines. They have "agents" located at the busiest and noisiest stations, whose job is to alert pilots by radio when they are in the vicinity of the box during takeoff. When the agent signals "mark," the pilot reduces power and maintains his altitude for a period of about ten seconds. Then, again on radio signal from the ground agent, the pilot resumes his normal noisy flight operation. The agent has done his job well, at the expense of the community. During September 1967, approximately 1,686 aircraft noise violations were illegally avoided this way. Because of such practices, only 690 official noise violations were recorded among 98,443 flights in the first nine months of 1967.[93] This procedure is still being used, and the FAA has not intervened.[94]

There are more subtle schemes used by the aircraft industry. The Airport Operators Council, Inc. (AOCI) offers a "solution" to the problem of introducing aircraft to a community which has never before experienced jet services. (This "solution" may well be of interest to those persons living in the 800 communities projected to have newly constructed airports by the year 1980.)[95] It is basically a public relations approach. The AOCI draws an analogy between the elderly spinster who does not like to be driven fast in an automobile and the jet plane first introduced to a new airport community. In time, if broken in slowly, the community, like the spinster lady, will happily begin to tell friends how fast the trip was made, and what a careful driver she had. As stated in their report to AOCI members,

> The airlines bringing jets into the community, with the understanding and cooperation of the pilots, can take these steps to make the introduction a pleasure instead of a problem:
>
> First, the community must become gently acquainted with the new noise. When jets are first heard they are awesome. The sound is different, the airplane seems lower (because it is bigger) and an element of fear is associated with the new noise. The usual introductory VIP flights are helpful in this respect because they are usually lightly loaded, carefully flown, and press stories about the "new service" accompanies them.
>
> More attention focused on the potential trouble areas at this stage should be productive. The airlines could include as VIP's, not just the leading political and business figures, but also the community, church and educational leaders from the potentially sensitive areas. This, together with appropriate emphasis in the press as to the benefits to the community the new service will bring should help get the old lady in the car and the car out of the driveway—to use our earlier illustration.
>
> Secondly, the introduction of scheduled jet service should be slow and cautious—both as to numbers of flights, loading of flights and operation

of flights. Put bluntly, if it's necessary to operate every flight with 50 percent fuel load for a while, then that's the way it should be operated until we "get the old lady and the car out of town." Pilots, station managers and dispatchers should *know, as a matter of company policy,* that they are expected to use every precaution possible to avoid creating noise annoyance to the community. They should *know* that the company *expects* and *intends* to take economic penalties for a while—a month, two months or more, if necessary—to be sure that the community *knows* that

(a) a jet aircraft isn't much different from other airplanes so far as they are concerned, and

(b) the new noise is just as safe as the old one.

By judicious testing of community reaction, airlines can readily determine whether in fact they have the car "out of town" yet. Continued interest in the community's welfare should be manifested in a genuine, conscientious manner.

Third, when it is clear that the community is used to the new noise to the extent that it is no longer a subject of conversation or fear, then gradually heavier loads might be tested. Over a prolonged period of time loads could be increased (if route segments and payloads warrant it) to the "maximum speed limit."

Fourth, when the "maximum speed limit" is reached it should be scrupulously observed. But, what is the "maximum speed limit?" It is the level of noise acceptable to the community. It might be expressed in absolute terms (such as "decibels" or "perceived noise decibels") or it may be more subjective. It may also vary somewhat from city to city and from community to community in relation to various runways at a given airport. It will be discernible at the local level even without absolute numbers by careful analysis of public relations—a continuous and careful testing of public sentiment as expressed at church socials, cocktail parties, civic association meetings, PTA's, Kiwanis, letters to the editor, press stories, and obviously in the City Council or the Courts if things have gone that far![96]

Some of this advice is included in the Airport Operators Aircraft Noise Kit, also prepared by the AOCI.[97] More detailed advice is offered on how to cope with noise trouble that may arise from the community. Much of it is thoughtful, but some of it is an outright attempt to manage people and manipulate the community in order to remove social opposition—opposition to aircraft noise that may affect the environmental health of a community.

Airline advertising has done its best to minimize negative aspects of operating aircraft by focusing on the passenger, not the community. The alleged amenity of the quiet afforded to the passenger is consistently stressed by Madison Avenue. Eastern Airlines proudly herald their fleet as "whisper jet" quiet.* If there is any doubt in anybody's mind, the two words are lettered on every airliner to prove it. McDonnell-Douglas describe the DC-9 as "quick and quiet" rapid transit, for they have placed "twin fanjets at the rear, to give you smoother quieter flights."* According to British Overseas Airways Corporation (BOAC), beginning in 1964 "things quieted down over the Atlantic." This happened because the engines were placed "back by the tail instead of on the wings,"* consequently the rear-mounted engines "leave all the noise behind." As a result of acoustical engineering improvements, the sound level inside the BOAC VC10 is portrayed as being "as quiet as an English meadow on Sunday afternoon" and "as quiet as a lagoon."* Compared to all other present aircraft BOAC purports to have "the quietest cabin in the air."*

Nearly all aircraft advertising is passenger oriented and the inference is that what is good for the passenger is good for the community. But, what about the community? It rarely benefits. Acoustical improvements are usually limited to cabin interiors. Present-day aircraft simply do not offer the same whisper jet quiet to communities as they allegedly offer passengers. Unlike passenger comfort, "community comfort" is never mentioned, even though by 1975 "15 million people will be living near enough to airports to be subjected to intense aircraft noise."[98]

However, with the introduction of a new generation of larger aircraft, advertisements now admit that the noise levels of current aircraft are offensive to airport community residents. "Neighbors of airports don't have to be told what's up. It's noise. But quieter engines are coming. And those engines will be quieter and virtually smoke-free."† In announcing their entry into the airbus race, Lockheed Aircraft Corporation for the first time claims to offer "quieter power" to both passenger and community.

> Three mighty Rolls-Royce high-bypass turbofan engines will make the quietest "sh-h-h" in the sky. Take-off and landing noise will be far below present jet levels, making this airliner a quieter, better neighbor for people on the ground as well as those in the air. In 1971, the low-noise Lockheed 1001 will start flying for many great airlines.†

It is further described as the quietest plane ever airborne: "Rolls-Royce's new three-shaft engines are the smoothest and quietest that ever flew."† The other major entry, McDonnell-Douglas DC-10, has also mentioned acoustical improvements over present generation aircraft. However, General Electric, manufacturers of the DC-10 engine, is less

---

*For this and other highly colored advertising language quoted below, see Appendix B.
†See Appendix B.

boastful than Lockheed. They advertise, "We have already succeeded in making the DC-10 engine quieter than the engines now powering most of today's jet planes."*

Communities will shortly be able to judge for themselves the accuracy of these words. If they do not prove true we may see unprecedented litigation. One can only hope that such advertisements are not found to be misleading the public.

Another area where industry often dissociates itself from any community responsibility is in establishing noise standards for its products. Whenever standards are developed, it is the industrial trade associations themselves that prepare the majority of them. In the absence of government control, performance standards concerning noise emission are thus written by business for business. The primary function of any trade association is to maintain the most favorable climate for business. The usual philosophy, then, behind writing a standard is primarily to set one which all members of the association can meet and only secondarily to benefit the consumer. When a specific decibel level is included in the standard, the normal procedure is to "shoot high." If "outside pressures" develop *later,* once a commodity is on the market, then there is a comfortable margin to reduce the noise it generates without major expense to association members.

There is evidence that the lawnmower industry, represented by SAE (Society of Automotive Engineers), used this tactic in writing an industry-wide regulation establishing noise emission limits for their power mower equipment. Although the lawnmower manufacturers could meet a standard by which the noise level from lawnmowers would not exceed 59 dBA at a distance of 50 feet,[99] they have not done so. Still, the members approved a more lenient standard of 72 dBA at the same distance. These companies felt no need to place an additional financial burden on themselves to quiet their product as long as the public was essentially content. A quiet lawnmower could be built for about $15 more than a conventional model. Here, as in aircraft advertising, mass media advertising aims to create the impression of noiselessness. One power mower manufacturer has a model called "the Peacemaker." His advertising copy reads: "designed to keep peace in the family and the neighborhood. Quiet as a church mouse."*

Power lawnmowers are far from quiet. As a group they easily exceed their own 72 dBA recommended guideline. In an investigation of 15 mowers having 4-cycle engines the average noise level was 97 dBC, with a range from 92 dBC to 105 dBC.[100] Audiometric examination

POWER
LAWNMOWER
INDUSTRY

---

*See Appendix B.

of 15 subjects after a 45-minute exposure indicated that some temporary threshold shift (TTS) in hearing usually occured. This shift, although not permanent, registered as high as 35 decibels, and nearly half of these subjects reported some tinnitus, or ringing in their ears. Another investigation by the Public Health Service concludes that exposure up to one hour per day, five days per week is safe, but "short exposures to lawnmower noise can cause over-stimulation of the ears with resultant temporary shifts in hearing."[101]

AIR CONDITIONING
INDUSTRY

The noise of air-conditioning units is another source of community annoyance. As a guide to communities contemplating legislation, the Air Conditioning and Refrigeration Institute has prepared a performance standard for air-conditioning noise. The Institute, supported by industry, recommends a 60 dBA noise limit measured at the closest property line.[102] In a random examination of both central and room air-conditioning units it has been found that nearly one third of the central type installations exceeded even this industrially determined standard.[103] Commenting on the heating, ventilating, and air-conditioning systems currently available to the public, a federal task force on the technical aspects of the noise problem has observed:

> . . . The current economic trend of many heating, ventilating and air-conditioning manufacturers toward designing high-velocity pressure equipment, thin wall ducting and the use of false ceiling plenum spaces and open corridors as return ducts has resulted in the creation of some of the noisiest systems in existence . . .
>
> Most major manufacturers claim there is no real demand for quieter products when their cost is made known to the prospective buyer. In other cases the consulting engineer may reject a higher priced unit that the owner might be willing to buy. There are indications that the cost problem arises because many manufacturers attempt to recover immediately a large share of development costs. This is an unreasonable burden on the initial purchasers.[104]

There are ways of getting around this 60 dBA limit without obstensibly attenuating the noise. A well-known manufacturer has simply altered the frequency distribution of the air-conditioner noise. Instead of permitting these units to generate noise within the mid-frequency ranges (frequencies used in determining decibel readings on the A scale), it has redistributed the spectral energy of the noise. Most of the sound energy now occurs in the lower frequency ranges, less sensitive to the A scale. This manufacturer has succeeded in meeting a recommended standard. But, since there is no basic noise reduction, the user

*See Appendix B.

(which includes the community) has not benefited at all. The noise signal remains as annoying as before.

Every major air-conditioning manufacturer constantly tells the consumer that its product operates quietly. Much like aircraft advertisements, air-conditioning ads emphasize purchaser comfort. The community, which continuously receives air-conditioning din during the summer months, is rarely mentioned. One manufacturer claims to offer a window unit model that "roars like a mouse,"* while another is presented as being "noiseless."* Frigidaire describes their Prestige Model (AEP-8MN) as "pin-drop quiet,"* the epitome of noise suppression. Similar claims have been made by General Electric, Westinghouse, Whirlpool, and Lennox. While indoors, individuals may well be able to converse, relax, and possibly sleep with these air conditioners in operation; outdoors the situation is different.

In introducing the 1969 air conditioners, some manufacturers began to mention the outdoor aspect of their operation. The Carrier Company, which had run the longest advertising campaign emphasizing indoor quiet, began mentioning outdoor quiet:

> Being round, it's more attractive. Compact. Efficient. Quiet. Very considerate of your neighbors, your shrubbery. . . . Heat and sound are blown straight up. Away from shrubs. Away from people.*

York Air Conditioning depicted their whole-house air-conditioning system as running "so smoothly, so quietly, your neighbors will never *hear* how comfortable you are."* The engineering noise controls were also briefly mentioned: "moving parts are isolated in a sound deadening chamber. The powerful motor and fan run slowly."* A subsidiary of Carrier also turned its attention to outside performance:

> Payne's all-new Remote air-conditioner will snug right up against the house and hide behind the roses and run without rattling and without waking up the neighbors and withstand nasty weather and send the sound up and away and help keep the roses from wilting and the dog from being so grouchy.*

This theme of community concern has continued in subsequent advertising campaigns. The noise output of the 1970 Carrier air-conditioning systems is referred to as a "neighborly kind of quiet."* Furthermore, a larger portion of advertising copy is being used for discussing noise reduction.

> More coil surface is exposed to outside air for quieter, more efficient cooling. The fan, a special design, moves more air quietly . . . It blows the

---

*See Appendix B.

heat and sound up, away from your flowers and neighbors. The hermetic compressor is nestled deep inside the unit for more quiet.*

Despite all these claims about noise-free operations, present day air-conditioning systems are not functioning quietly in the outdoor environment. Collectively these manufacturers do not meet their own self-imposed 60 dBA noise limit.[105] Community noise ordinances like the one adopted by Coral Gables, Florida, are finding that most residential air conditioners exceed the maximum allowable noise level.

AUTOMOBILE
INDUSTRY

The automobile industry advertises quiet more than any other industry. Ford Motor Company was the first American car manufacturer to use the "quiet theme" nationally.† The earliest advertisements compared the Ford to the Rolls-Royce, with Ford offering a quieter ride. They still make reference to this earlier comparison: "LTD by Ford is designed to ride quieter than the Ford LTD that was quieter than a Rolls-Royce."* Beginning with the introduction of the 1969 automobiles, other manufacturers also began mentioning noise control. However, the 1970 model year emphasized the amenity of quiet more than any previous year. On all of its 1970 cars, Chrysler Corporation offered a "Torsion-Quiet Ride" described as "a unique suspension system that insulates against road shock and engine noise."* Ford Motor recommended that the consumer should "Take a Quiet Break in the 1970 Ford. We don't just cover up the noise. We build in the quiet."* Two divisions of General Motors also launched "quiet car" campaigns. The 1970 Chevrolet Monte Carlo was publicized as being "sailplane silent," while the new Cadillac V-8 engine "smoothly and quietly delivers a responsiveness that's astonishing for a car of such magnificence."*

With the newest crop of automobiles it is unusual when a manufacturer fails to mention the subject of acoustics. Even some imports, which have been generally noisier than their American counterpart, now are following suit.

The primary emphasis is silencing noise inside the car; passenger quiet rather than community quiet receives the attention. Ford typifies the present engineering emphasis:

The quiet begins deep. With all systems working together to silence, hush, absorb. A computer designed safety frame has integral torque boxes to swallow hush vibration. . . . Inside, you relax and enjoy the quiet, in

---

*See Appendix B.
†The Dorris Motor Car Company incorporated the theme of quiet in their advertising on a regional basis only. (See p. v.)

padded, panelled luxury. . . . And new ventilation system so quiet only the coolness tells you its working. Isn't this the way driving should be in this noisy, nerve-wracking world?*

Contributing to this "noisy, nerve-wracking world" is the automobile, which is not nearly as quiet outside as inside.

What about the inside? Is it as quiet as most car manufacturers lead the consumer to believe? The answer is a resounding no (see Figure 1.2). I have compared advertisements used to describe 1970–1971 cars with actual test results performed by *Popular Science* personnel.[106] The use of the word "quiet" is grossly misused. There is a 16 decibel variation between the "quiet" the 1970 Ford Thunderbird (67 dBA) offers the buyer and the "quiet" offered by the 1971 Chevrolet Vega (83 dBA). Furthermore, there is no correlation between the automobile's price or size and the level of noise a buyer should expect. Many large and expensive automobiles were as noisy as the smaller, less expensive automobiles. It is also evident that no American manufacturer produces a complete line of quiet cars. For example, General Motors manufactured a quiet Chevrolet Monte Carlo, but at the same time was responsible for a very noisy Chevrolet Vega. The same dichotomy exists at Ford Motor Company with the Ford Thunderbird on the one hand and a comparatively noisy Ford Mustang on the other, and also at Chrysler.

The question of credibility must be raised. Automobile advertising is conveying misinformation. It should not be allowed to continue unless quiet, silence, or any other synonym, is defined in decibels, by, or, if necessary, for the manufacturer. Manufacturers of automobiles as well as of other consumer goods that advertise their products in this manner should be required to validate their acoustical assertions. Two consumer organizations with Ralph Nader have filed a petition with the Federal Trade Commission that could be applicable in this area. They proposed that the FTC require national advertisers "to file with the commission evidence substantiating their statements at the time the advertisements were published or printed."[107] This information would then be made available to the public.

SNOWMOBILES

Up until the 1960's, our wintry landscape offered the opportunity to escape the noise of urban living. This decade brought with it an ecologically devastating device that is transforming the tranquil countryside into a noise-congested environment. Although the snowmobile had modest beginnings, with sales during 1963–1964 of no

*See Appendix B.

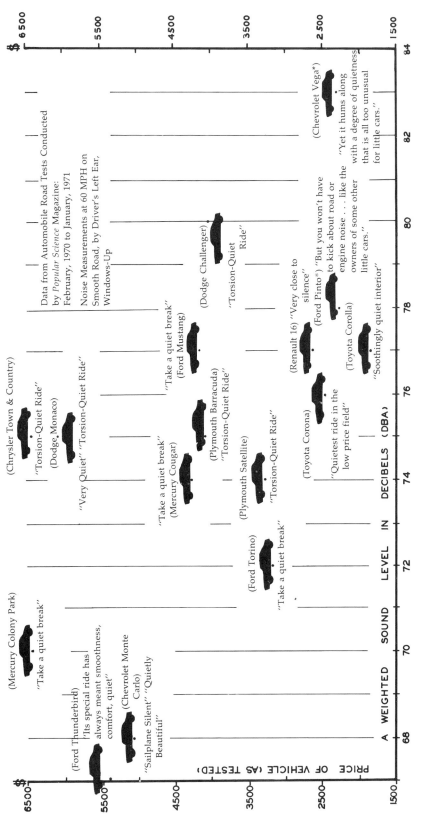

FIGURE 1.2   *Automobile advertising and actual interior noise levels* (1970). *1971 models

more than 15,000,[108] they now are found wherever there is snow. By the end of 1971, the industry estimates, there will be over 1.5 million in use.[109]

Powered by some engines greater than 40 horsepower and capable of speeds over 60 miles per hour, the snowmobile is not only a noise annoyance problem, but also a very real noise hazard to driver and passenger. Sound pressure levels can register over 120 dBA on the more powerful models when operated at full-throttle. Sounds at these levels approach the threshold of pain.

Temporary and even permanent hearing loss are possibilities. Poynor and Bess have found exposure to snowmobiles can cause a temporary hearing loss ranging from four to fourteen days, depending upon the particular model.[110] Despite these reported hazards, the snow-mobile industry has shown little interest in protecting the hearing of its customers. It is only with the introduction of the 1971 models that their advertising has even mentioned the subject of noise. Johnson, a division of Outboard Marine Corporation, is the only one to my knowl-edge to do so: "The engine is completely enclosed. So is noise, dirt and ugliness." * But what about the majority of the over one million snow-mobiles operating in this country with little or no acoustical treatment? They are just one more noise source contributing to the deterioration of man's hearing. Most of the nineteen states having laws regulating snowmobile usage are now requiring that these vehicles not exceed 86 decibels.[111] This noise requirement however is based upon measure-ments taken at 50 feet from the snowmobile. Consequently the snow-mobile operator is still exposed to a potentially hazardous noise well above 86 decibels.

Aircraft, lawnmowers, air conditioners, automobiles, and snow-mobiles represent some of the more offensive community noises. Some manufacturers of these products realize a problem exists, and a few are trying to correct the deficiency. Other industries have either avoided the problem completely or have treated it superficially. Major segments of private industry are directly responsible; they have not allocated the necessary dollars for engineering out product noise. One of the largest power tool companies and one of the largest tire manufacturers do not have acoustical instrumentation necessary for evaluating, much less reducing the associated product noise. All testing and evaluation is reliant upon human judgments. The public deserves the right to choose quieted products.

The consumers' patience is wearing thin, and consumer-protective legislation may well deal with the noise menace directly. Meantime, the public interest has received some protection through an amendment to the Walsh-Healey Public Contracts Act, enacted May, 1969.[112] This act now establishes occupational noise thresholds which could make

*See Appendix B.

industry think more seriously about controlling noise in general. The U.S. Occupational Safety and Health Act and Standards (adopted May, 1971) establish mandatory occupational noise thresholds for all businesses in interstate commerce. Also, nuisance litigation has affirmed the principle of product liability which asserts that the manufacturer of a consumer item is liable for any hazards to which it exposes the consumer. One of the most vocal spokesmen from industry has reminded negligent companies of their obligation to abate noise they themselves create.

> The manufacturer is obligated to incorporate safety features into his product that will reduce the possibility of injury to the ultimate user. Legal precedents involving product liability affirm this principle. Excessive noise is a hazardous attribute of many products that manufacturers have not given the corrective attention they deserve.[113]

Conflicts in Societal Goals

Evidence is mounting that some primary goals of urban living, such as the preservation of health and environmental well-being, are being sacrificed to secondary goals such as economic growth and technological development. Secondary goals ought to serve the achievement of health and societal well-being. But when technology, aiming solely at economic growth, pays little attention to environmental hazards and nuisances, the primary goals become the victims for "progress." A healthy Gross National Product does not insure there is also a healthy environment in which to live. In fact, some believe they are diametrically opposed, that is, environmental quality declines as the GNP advances.

We can ill afford to accept blindly technology's present direction while disregarding its effects. Yet many of us are doing exactly this. Environmental noise pollution is a direct by-product of our technologically oriented way of life. But it is not an *inevitable* price which must be paid for so-called "progress." Once noise is considered inevitable, we have lost all objectivity. Health and environmental well-being then become subservient to economic expediency. The environment then is allowed to remain noisy and unchanged.

At the federal level, where there is considerable responsibility for noise control, certain goals often receive preference at the expense of objectivity. The Federal Aviation Agency (under the Department of Transportation) exemplifies this conflict and compromising of societal goals. The fault lies partly with Congress and the enabling legislation creating this regulatory agency in 1958. Under the provisions of that act, the FAA has several purposes, including:

1) encouraging the use and development of the aircraft industry
2) providing safety and welfare to the public.[114]

The development of air transport is presumed to be desirable in terms of its positive impact on economic growth and consequent provision of jobs. But little consideration has been given to the substantial side effects which arise therefrom. Technological and economic development of the aircraft industry has come about, but at the public's expense. Aircraft noise remains basically unchecked. Within the past ten years, as the industry has rapidly expanded, there has also been a "steady decline in habitability of residential areas near the large airports, areas containing tens of thousands of families, their schools, and their hospitals."[115]

As a result, strong public discontent has arisen in communities close to airport facilities. The community's concern is being brought more and more urgently to the attention of government officials. General William F. McKee, while he was FAA director, said: "Noise means irritated citizens whose growing protests are blocking needed airport expansion even when such money is available."[116] Former Assistant Secretary of Transportation Cecil M. Mackey went further, saying that the citizen's insistence on less degradation of his environment "is the single most outstanding characteristic of society."[117] He warned that if this insistence cannot be answered with acceptable solutions to aircraft pollution, congestion, and noise, "people will just say 'Sorry, we don't want airplanes around any more.'"[118] Community resistance to the construction of new jetports is mounting across the country, and noise is one of the primary reasons for this resistance. Currently there are approximately 50 of our 140 major airports involved with noise litigation[119] having claims for damage approaching $4 billion.

The FAA has not consistently supported the citizen's position. Key personnel in the FAA and in the Department of Transportation, according to reports of the Conservation Foundation, "have indicated in the past that aviation noise is not one of their primary concerns."[120] Economic development of the aircraft industry has been a preference by DOT officials. Former Secretary of Transportation Allan S. Boyd earlier commented, "Noise is a very unfortunate and disturbing thing, but we do learn there is room for more tolerance of noise in the field of aircraft."[121] A similar position is shared by many current departmental officials.

In the past, the Department of Transportation has also been insensitive to vocal community complaints. When he was chairman of the Civil Aeronautics Board, William Boyd suggested that the government should get the advice of psychologists on how to deal with community noise protests which could probably be traced back to the "anxiety psychosis that seems to dwell over a great many people nowadays."[122]

In many places, public accommodation to noise around airports has already gone beyond the tolerance point, and the economic rationalization does not console the affected population. Congressman Byron Rogers has raised the basic question:

> . . . How far and under what conditions can the community impose the consequences of technological progress on individuals? We know how economically important is the aircraft industry. Stapleton International Airport in Denver provides employment for more than 6,500 persons at an annual payroll of more than $55 million, yet such an economic argument has no effect or weight on those annoyed by unreasonable noise . . .[123]

The technology creating subsonic aviation noise and community annoyance may begin to offer the necessary solutions—under pressure from the public law establishing engine noise maximums for certain aircraft.[124] If this public law is subsequently followed by noise certification procedures for helicopters, VTOL, and STOL, and an engine retrofitting requirement on aircraft now in use, a major effort will have been made to balance two of FAA's seemingly contradictory purposes —protecting the public while encouraging the use and development of the aircraft industry. Without adequate solutions to effectively control subsonic aircraft noise, we began the process of creating a more pervasive noise source, the supersonic transport and became more and more committed to its success as our entry in the SST race. Major congressional opposition appeared when the program was in development Phase 3, which entailed constructing two prototypes. Nearly $800 million had been approved by Congress, and at the end of this phase the federal government's share was to have been 90 percent, or $1 billion, of the development costs. While federal appropriations continued to be voted annually, concerted congressional opposition grew.

This program ran into its first concerted opposition during the 91st Session of Congress. The House of Representatives narrowly approved a $290 million SST appropriation for fiscal year 1971 (176 for and 163 against). But, on December 3, 1970, the Senate rejected the amendment to continue funding the project. It was defeated 52 to 41.[125] When the House members refused to defer to the Senate vote, a House-Senate committee was appointed to work out a compromise agreement. This committee gave a caution light to the SST program by recommending reducing the appropriation by $80 million, down to $210 million. Although the House was satisfied, the Senate as a whole rejected this compromise. SST opponents in the Senate called this a sell-out and Senator William Proxmire threatened to lead a filibuster until the end of the session if necessary.[126] SST proponents led by Senators Warren Magnuson and Henry Jackson failed twice in an attempt to invoke

FIGURE 1.3. *Airborne industrial noise*

the people know the problem, but what to do about it is usually entirely beyond their knowledge. Nearly every Tinicum resident we interviewed considered aircraft the principal noise problem, but when asked how the situation could be improved, responses were: "What can you do about it anyway?" "You can't fight City Hall," "You can't stop progress," "I only wish I knew," or "The only answer is to move." These opinions are probably typical for most Americans.

Every day, more people settle in housing near noise sources such as airports, freeways, and commuter lines, without recognizing the magnitude of the noise problem. In Rosedale, Long Island, close to JFK International Airport, a plan for a multiple-family residential development was approved. The developer, who had previous success in the same area, told New York City officials that when he built "Country Estates" (a single-family development), people knew of the proximity of the airport and were aware of a potential noise problem.[147] However, this fact proved to be no obstacle in selling the houses. These houses ranged in price from $37,000 to $48,000, "plus options," and incorporated no sound insulation materials.[148] Multiple family apartments are continuously rising close to the busier airports where high noise levels are never-ending. The "Twin Pines" in New York and the "Lamp Lighter Towers" in Chicago, situated in the worst possible locations relative to flyover noise, have occupant waiting lists. (But as one city planner has commented, Chicago area housing is so tight right now, you could build an outhouse at the end of a runway and someone would rent it.)

PUBLIC NOISE
ABATEMENT
ORGANIZATIONS

There are very few avenues through which the public can learn about and defend itself against either the potential hazards of noise or about what can be done to suppress them. There is no longer a nationally based organization concerned with noise pollution. The National Noise Abatement Council (NNAC), once active, disbanded in 1961. This organization had sponsored yearly noise abatement conferences, but those companies supporting the Council withdrew because interest declined. NNAC also had presented awards to cities that had done the most significant abatement work. Philadelphia, for example, won the award in three consecutive years, 1953–1955. The Council also disseminated literature to interested parties about the noise problem and about what could be done to make our cities more habitable, but it "never achieved the prominence or effectiveness that the ostensible public outcry against noise should have warranted."[149]

After a six-year hiatus, two noise abatement organizations started to represent the public's interest once again. The Citizen's League Against the Sonic Boom (CLASB) started operations in 1967 under the direction of Dr. William Shurcliff, a Harvard University physicist. By the end of that year their membership had reached 2,200.[150] Today numbering well over 3,000 members, the league is carrying on a campaign opposing the development of the SST in conjunction with other conservation groups. A newspaper available to CLASB members reports the status of the U.S., Anglo-French, and Russian SST projects. Their activities have included sponsoring national advertising, directing a congressional write-in campaign, as well as preparing a "S/S/T and Sonic Boom Handbook," for members of Congress, governmental officials, and journalists. This same handbook has now been published in paperback by Ballantine Books.[151]

Citizens for a Quieter City, Inc. (CQC) was organized in 1966. This organization is concerned about urban noise in general, with its geographical focus being New York City. Its executive director, Alex Baron, is convinced that "education and applied research are prerequisites to intelligent action." Consequently, the purpose of Citizens for a Quieter City is

> to create an awareness of the need for the control of city noise and of the means available for such control. We want to stimulate research not only on hardware and legislation, but on the effects of noise on man.[152]

As a citizens' group the members have lobbied and supported efforts to limit the use of heliports and short takeoff landing aircraft (STOL) facilities in residential parts of Manhattan, to reduce the din of refuse trucks by including noise performance specifications on city contracts, to pass the New York City building code with its noise provisions, and to acquaint the public with acoustically treated products available on the market (i.e., quieted air compressor, pavement breaker, and refuse cans). Under Baron's leadership the organization has also sponsored forums on noise pollution and the need for control. A recent undertaking has been a public advertisement project urging the public to recognize that "the citizenry has at least an implicit right to a reasonable amount of quiet—and present technology is capable of insuring that right." One full-page anti-noise advertisement simply states "Noise pollution won't kill you. It can only drive you nuts or make you deaf." While doing a commendable job of supporting the public's interest, this organization has limited financial resources, and is designed to assist primarily the New York metropolitan area. CQC has been awarded a $300,000 Ford Foundation grant to examine methods for reducing noise in a 60-block area of New York City.[153]

Their proposed solutions should be applicable to other noise-inflicted communities. Beside CLASB and CQC, there are few agencies from which the citizen can learn about noise and how it can be controlled.

One potential federal resource is the Consumer Protection and Environmental Health Service. Consumer problems are now being examined by this service, as well as by several congressional committees. The objects of these examinations include automobile warrantees and consumer service, product misrepresentation, and consumer packaging and labeling. No consumer education and protection program dealing with noise has as yet been launched, but such a proposal has been submitted to the Department of HEW.[154] The closest thing to a federal education program was the work done by former Congressman John Kupferman from New York. When he was a member of the House of Representatives (1968–1969), his office acted as a national clearinghouse on noise matters. Periodically, Representative Kupferman had noteworthy articles and congressional correspondence reproduced in the *Congressional Record.* For a while after that Senator Mark Hatfield reported on the subject. (A very important contribution was a compilation of state and local ordinances on noise control the Senator had printed in the October 29, 1969, issue of the *Congressional Record.)*[155] However, the practice has not been routinely continued. The general public has rarely become aware of this educational material for other reasons too. Readership of the *Record* is neither large nor diversified, an annual subscription price of $45.00 is now in effect, and there is no longer a subject index to expedite its reading.

PUBLIC INTEREST
GROUPS/
ORGANIZATIONS

Several conservation groups are becoming more and more in-terested in noise as part of the total environmental pollution package. Their primary involvement to date has been in opposing the SST, and any further federal expenditures for it. Referred to as the environmental coalition, the Sierra Club, the National Wildlife Federation, the National Wilderness Society, and Friends of the Earth with CLASB engineered a very effective congressional lobby to curtail, if not eliminate, federal support for the SST program.[156]

During the 1970 U.S. congressional campaign the League of Conservation Voters also moved into the noise area. They compiled a congressional voter list on ten key environmental issues, including the House of Representatives vote on the SST appropriations (HR Bill 17755).[157] This listing was distributed widely. Those candidates found to be the most conservation oriented, on the basis of their voting record, were endorsed and actively supported by the League.

Toy safety is another area where the hazards of noise have been discussed. Consumers Union played a major role in the passage of the 1969 Toy Safety and Children Protection Act.[158] Under the emergency provisions of this act, they filed a petition in November 1970 with the Secretary of Health, Education and Welfare demanding that the Secretary declare nine toys as "imminent hazards to the public health." This list contained a cap pistol that "made such a loud bang that deafness or substantial loss of hearing could result," according to Consumers Union's technical director.[159] Laboratory tests demonstrated that the peak impulse noise level exceeded 155 decibels, one foot in front of the muzzle.[160] (Peak impulses exceeding 140 decibels generally are considered hazardous to hearing.) As early as 1966 there was evidence that impulsive noises produced by certain firecrackers and toy firearms were acoustically hazardous to children.[161] Consumers Union and the Children's Foundation sought with their petition to force these toys off the market before the 1970 Christmas shopping season. These two groups subsequently filed an injunction with the federal district court in early December, after the Department of HEW failed to act. With just three shopping days left before Christmas, the Food and Drug Administration finally banned 39 toys as dangerous, requiring their removal from all retail outlets.[162]

SCIENTIFIC
INFORMATION

Professionalization has further restricted the vital flow of scientifically based noise information to the public. Acoustical engineers, industrial hygienists, behaviorists, and other professionals appear to guard their interests in this pervasive problem by writing in a jargonistic, "professional" manner. Only occasionally does the public receive translations through the wire services or popular magazines. Usually these are watered down and incomplete. For eight years, the Acoustical Society of America supported a very competent journal on the subject entitled *Noise Control*. Because of certain economic considerations their sponsorship was terminated in 1961, and the highly readable magazine fell by the wayside.[163] Although the Acoustical Society assured readers of *Noise Control* that the *ASA Journal* would incorporate the subject of noise, it does not often publish community noise articles.[164] The articles that do appear are usually highly technical.

Fortunately there still are some links, if only a few, between the scientific community and interested citizenry. One such link is a monthly magazine entitled *Sound and Vibration*. It has a growing readership of over 12,000. All aspects of noise are reported. Topics cover evaluation of various noise sources, engineering methods for controlling noise, auditory and nonauditory effects of noise, noise control parameters for industry and community, as well as news in the field of acoustics.

Recently it instituted a series of noise control buyer's guides. Each guide lists the product, manufacturer's name, and address. The series includes guides on: "Sound Absorption, Sound Barrier, Vibration Isolation and Vibration Damping Materials";[165] "Mufflers, Sound Absorption Systems, Sound Barrier Systems and Vibration Isolation Systems";[166] "Hearing Protective Devices";[167] and "Dynamic Measurement Instrumentation Buyer's Guide".[168] This is the first time a listing like this has ever been compiled for a general readership.

Sound and Vibration does have a few shortcomings, but they are minor. There is a controlled circulation policy; however, "management, engineering, scientific, or technical personnel—in industry, government, or education—having a responsible interest in the control, measurement or generation of sound and vibration" receive the magazine gratis.[169] Those people not meeting these requirements can subscribe at ten dollars per year, though basically it is read by professionals.

There are two more widely read magazines that appear to be educating the consumer in this subject. *Popular Science* magazine now includes interior noise level measurements as part of their automobile road tests (see Table 1.4). Begun in February 1970, the interior noise measurements are recorded under three sets of conditions (30 and 60 miles per hour on smooth road, and 30 miles per hour on rough road), using three different weighting networks—sound level in dBA, dBB, and dBC.[170] (Recent road tests report the decibel level using only the A-weighted scale, or dBA.) A General Radio sound level meter is positioned inside the car by the driver's left ear, with the windows completely closed. Since these road tests are all performed on the same

TABLE 1.4   Interior Noise Levels: 1970 Automobiles

Automobile	30 miles per hour on smooth road	60 miles per hour on smooth road	30 miles per hour on rough road
AMC: Gremlin	68 dBA	77 dBA	75 dBA
AMC: Hornet	67	74	78
AMC: Javelin SST	65	76	76
AMC: Rebel	63	75	74
Audi 11-LS	62	76	73
Buick Estate Wagon	61	73	68
Buick Riviera	59	68	68
Chevrolet Camaro SS	64	71	74
Chevrolet Malibu	63	69	71
Chevrolet Monte Carlo	61	68	73
Chevrolet Suburban Carry-all	64	76	78
*Chevrolet Vega	68	83	82
Chrysler Town & Country	62	75	75
Citreon D Special	64	75	73
Datsun 510	68	77	77
Datsun 1200	68	79	72
Dodge Challenger 340	65	80	75
Dodge Monaco Wagon	62	75	77
Fiat 124S	67	79	76
Ford Mustang Mach 1	67	77	78
*Ford Pinto	67	78	84
Ford Thunderbird	58	67	71
Ford Torino	63	72	74
International Travelall D-1000	67	71	71
Kaiser Jeep Wagoneer	65	73	78
Mercury Colony Park	60	70	69
Mercury Cougar Eliminator	64	75	74
Oldsmobile Toronado	59	72	69
Peugot 504	62	78	70
Plymouth Barracuda 340	66	75	75
Plymouth Satellite	64	74	77
Pontiac Firebird 400	67	74	74
Pontiac Grand Prix	60	68	72
Renault 16	64	77	74
Saab 96	68	77	78
Simca 1204	73	80	78
Toyota Corolla	66	77	80
Toyota Corona	66	76	76
Volkswagen 1600	72	80	82
Volkswagen Squareback	67	81	78
Volvo 144S	64	78	72

Source: Automobile Road Tests Conducted by *Popular Science* personnel and reported in *Popular Science* between February 1970 and January 1971.
*1971 model.

test facility, using identical test procedures, the results offer the potential car buyer valuable comparisons. Some thirty-nine 1970 foreign and domestic cars were tested, with a decibel range of 67 dBA to 84 dBA at 60 m.p.h.! Unfortunately they do not perform noise measurements on any other consumer items like the noise hazardous snowmobile. Other helpful articles have included construction guidelines for erecting acoustically treated barriers to minimize sound transmission from room to room.[171]

Consumers Union, through *Consumer Reports,* is acquainting their members with the subject of product noise. In evaluating table saws, the magazine discussed recorded noise levels, and the potential hearing loss problem.[172] Their evaluation of children's toys uncovered a noise-hazardous cap gun, and a diesel tractor having an annoying noise maker that simulated engine sounds. Consumer articles such as these are performing an important role of educating the public to think and buy quiet. However, most products rated by Consumers Union are only judged as to their relative noisiness, and no sound level measurements are performed. Reporting on a particular product, they may state brand X is "relatively quiet," or brand Y is "excessively noisy." Without the actual data, the reader is given less definitive information upon which to formulate an opinion about a product. Furthermore, personal judgments can lead to possible errors.

In a test of six little cars, they reported the Chevrolet Vegas as the quietest tested, quieter than either the Ford Pinto or Toyota Corona.[173] (There was no mention of actual sound measurements, so it is assumed these were judged comparisons.) In contrast, *Popular Science,* with sound level measurements to support their position, found the Toyota decidedly quieter than either the Pinto or Vegas (see Table 1.5).[174] Hopefully, Consumers Union will consistently use sound measuring instrumentation as part of their product evaluation program, and these results will be published for the reader's benefit.

The general subject of noise is also receiving growing attention from the book publishing world. Last year brought the first wave of

Noise Level Comparisons: Automobiles   TABLE 1.5.

Automobile	Noise levels		
	Smooth road (30 mph)	Rough road (30 mph)	Smooth road (60 mph)
Toyota Corona	66 dBA	76 dBA	76 dBA
Ford Pinto	67 dBA	84 dBA	78 dBA
Chevrolet Vega	68 dBA	82 dBA	83 dBA

books oriented toward the reading public. *The Tyranny of Noise,* written by Robert A. Baron, gives a deep insight into the very public problem of urban noise.[175] Two others, *The Fight for Quiet,* by Theodore Berland,[176] and *In Quest of Quiet,* by Henry Stillman,[177] echo the general statement of the problem, though they offer solutions more as observers than as participants. Despite some technical errors, particularly in the latter two, these books are all very readable and fill a definite public information vacuum. The lack of public awareness still continues because the available information resources are inadequate.

We in the United States are losing our ability to judge our environment. That environment is growing noisier over ever-greater periods of the day; the periods of comparative quiet are diminishing. Since we live with it, noise has become an integral part of living, and quiet is more difficult to perceive. To a growing number of Americans, the absence of noise is annoying, since noise is a part of their daily diet. The younger, "transistorized" generation is "turned on" by noises of life. Even in Tinicum Township, where aircraft noise was most intense, several of those interviewed "missed" their noise when aviation stopped because of inclement weather. In a period of urban living where noisiness breeds noisiness, it is important to re-educate the population to the virtues of quiet.

What Is Noise?

What is meant by noise, and how can it be measured? It is easier to describe the physical properties of noise and its effects than to offer a precise definition of the term. However, for all practical purposes noise can be defined subjectively as unwanted sound, sound not desired by the recipient.[1]

The physical or acoustical measurement of noise involves understanding three characteristics of sound: intensity, frequency, and duration.

Characteristics and Physical Measurement of Noise

The physical measurement of a given sound is determined by measuring its pressure relative to a base or reference sound pressure. This difference indicates the intensity of a particular sound, or its sound pressure level (SPL). Decibels are dimensionless units used to describe sound intensity.

SOUND INTENSITY

The formula for determining the number of decibels a sound generates is:

$$SPL = 20 \log 10 \frac{P}{P_0} \, dB$$

SPL refers to the sound pressure level of a measured sound in units of decibels; dB or decibels express the logarithmic ratio of a measured sound pressure to a base sound pressure; P is the average pressure of a measured sound; while P_0 indicates the reference pressure considered to be the weakest audible pressure a young ear can detect under ideal listening conditions (0.0002 microbars).[2] In establishing the decibel scale, the reference pressure P_0 is assigned an intensity of OdB. Accompanying a more intense sound is a greater decibel level. The

TABLE 2.1.   Noise Levels Measured in Microbars

Sound pressure (microbars)	Sound pressure level (dB re 0.0002 microbar)	Noise source
0.0002	0	Threshold of hearing
0.00063	10	
0.002	20	
0.0063	30	
0.02	40	
0.063	50	
0.2	60	Conversation
0.63	70	
1.0	74	Vacuum cleaner
2.0	80	
6.3	90	Subway
20	100	Snowmobile
63	110	Air hammer
200	120	Chain saw
2000	140	22 caliber rifle

logarithmic component in the formula means that the geometric increase in sound pressure can be represented by an arithmetical increase in the decibel level. For every 10 dB rise in the scale, the pressure of a given sound more than triples (see Table 2.1).[3]

Since the decibel is a logarithm of a ratio of two values, it can also describe sound energy. Energy equivalents for a sample of noise sources are presented in Table 2.2. As shown with every arithmetical increase in decibels which is a net gain of ten (for example 10 dB to 20 dB, 20 dB to 30 dB, or 50 dB to 60 dB), the relative change in the sound energy increases geometrically ten times. On jumping twenty decibels from 60 dB to 80 dB, *the corresponding change in sound energy represents a rise of 100 times, or from 1,000,000 to 100,000,000.*

Some confusion may arise in calculating the overall noise level of multiple noise sources. For instance, the noise generated by two jet aircraft each making 120 dB is *not* 240 dB, the two added together.

TABLE 2.2.   Noise Levels Measured in Sound Energy

Relative change in sound energy	Decibels	Noise source
1	0	Threshold of hearing
1,000	30	Whispering
1,000,000	60	Conversation
100,000,000	80	Food blender
10,000,000,000	100	Heavy traffic
1,000,000,000,000	120	Jet aircraft

The actual decibel level is 123 dB, or slightly more than the level of one aircraft alone, but the energy level is doubled (0.3 = the log of 2). Numbers of decibels are never directly added; first decibels are converted to relative power, added or subtracted, and then converted back to corresponding decibels. The same procedure is required when combining unequal noise levels, such as a truck's noise (90 dB) and the noise of a motorcycle (88 dB). If these two are being combined, the total is not 178 dB, but rather 92.1 dB. A chart has been developed for combining levels of uncorrelated noise signals (see Figure 2.1).[4]

The combined noise of an elevated-subway car (90 dB) and a bus (84 dB) can be calculated by using this chart in the following manner:

(1) determine the difference in levels (6 dB)

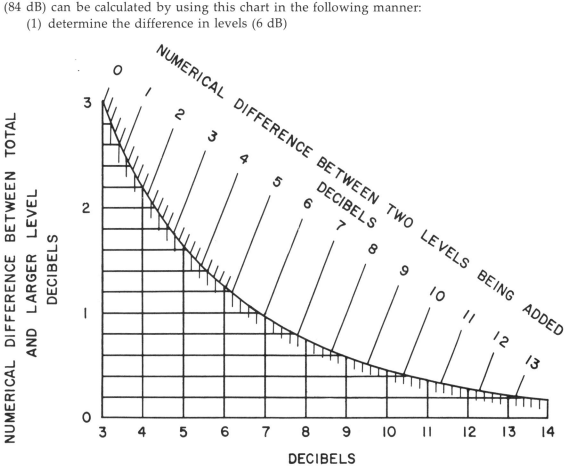

FIGURE 2.1. *Chart for combining levels of uncorrelated noise signals.* **Source:** *A. Peterson and E. Gross, eds.,* Handbook of Noise Control *(West Concord: General Radio Company, 1967).*

(2) locate 6 dB on the "To add" scale

(3) read directly across to the left hand vertical scale to the answer (approximately 1.0 dB)

(4) add this amount to the higher noise, the subway noise (90 dB + 1.0 dB = 91.0 dB)

The aggregate of these two noise sources is 91 dB.

FREQUENCY   To assess community noise, it is important to consider more than overall intensity, because noises of this type generally are not pure tones. Sound waves generated by sources of community noise consist

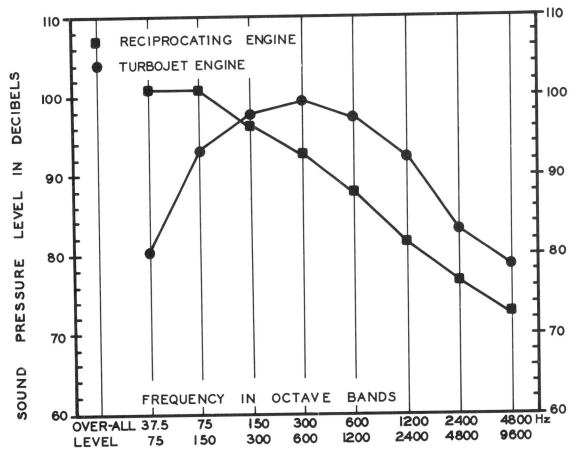

FIGURE 2.2.   *Frequency distribution of aircraft noise measured in cockpit. Data from Major Donald Gasaway, "Aeromedical significance of noise exposures associated with the operation of fixed- and rotary-winged aircraft" (USAF School of Aerospace Medicine, Brooks AFB, Texas, November, 1965, mimeographed).*

of a range of frequencies, from the low frequency roar of traffic to the high-pitched whine of jet aircraft. The frequency of sound represents the number of times a complete cycle, consisting first of an elevation and then of a depression below atmospheric pressure, occurs in one second.[5] Hertz (Hz) is designated as the unit for measuring frequencies.

A frequency analysis is used to determine the way in which noise intensity is distributed. The acoustical energy of a noise is electronically divided into passbands. Usually there are eight bands used, with each band being an octave. A sound pressure level is obtained for every frequency band. The eight commonly used bands are centered about the preferred frequencies for acoustical measurements which are 63 Hz, 125 Hz, 250 Hz, 500 Hz, 1,000 Hz, 2,000 Hz, 4,000 Hz, and 8,000 Hz.

Two noises having similar overall sound pressure levels can display completely different frequency distributions. An example is the sound of an airplane. A four-engine turbojet aircraft is compared to a single reciprocating engine aircraft in Figure 2.2.[6] Though they have similar overall sound pressure levels (104 dB and 107 dB respectively), their frequency distributions by octave bands vary markedly. With the jet, most of the energy is concentrated in the middle frequencies, which are irritating to the perceiver. In contrast, the propeller-driven plane registers the majority of its sound energy in the lower, less irritating frequencies below 300 Hz.

In subsequent chapters, references will be made to a sound level in decibels or "dB" followed by another letter A, B, or C. These letters correspond to specific frequency response curves of standard sound level meters that selectively discriminate against low and high frequencies (see Figure 2.3).[7] Each letter depicts a specific "weighting" network. The C-weighting network closely resembles the "flat response" curve, passing all the frequencies nearly equally. The other two networks (A and B) progressively attenuate or suppress frequencies below 1,000 Hz.

Of the three scales, the A-scale has become the most popular weighting network in community survey work, because its frequency response corresponds to the way the human ear perceives sound. Nearly all the noise analyses of the West Philadelphia community utilized this scale. Where reference is made to the A-weighted network it is indicated as "dBA," or the "A-weighted sound level in dB." When the sound level is referred to in decibels, or *dB alone* (without A, B, or C), then it is assumed that the *C scale* is being used.

The duration of a noise is a third consideration. Hearing loss, for example, is directly affected not only by intensity and frequency but also by the duration of exposure. The Second Intersociety Committee

DURATION OF NOISE

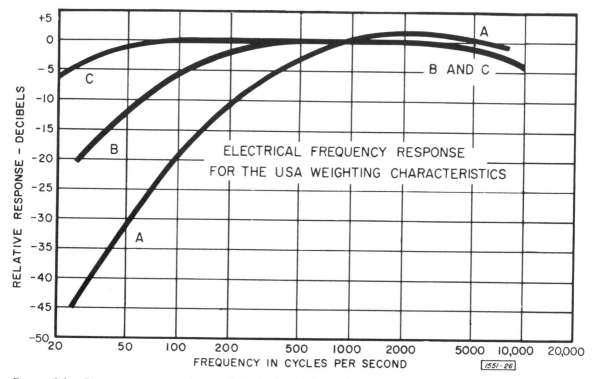

FIGURE 2.3.   *Frequency response characteristics in the ANSI standard for sound level meters.* Source: *A. Peterson and E. Gross,* Handbook of Noise Control *(West Concord: General Radio Company, 1967).*

on Guidelines for Noise Exposure Control* recommended maximum noise exposures as a function of time (see Figure 2.4).[8] This committee "considers that a reasonable objective for hearing conservation is an environmental level of 90 dBA for steady state noise with permissible increase of 5 dBA for each halving of the exposure time up to 115 dBA." Under this scheme, an individual can be exposed to a sound level of 105 dBA for 60 minutes, or for 30 minutes to a sound level of 110 dBA. Endorsed by the representatives from the American Council of Governmental Industrial Hygienists, American Academy of Ophthalmology and Otolaryngology, American Academy of Occupational Medicine,

---

*Committee of noise experts representing five professional organizations: American Council of Governmental Industrial Hygienists (ACGIH), American Academy of Ophthalmology and Otolaryngology (AAOO), American Academy of Occupational Medicine (AAOM), American Industrial Hygiene Association (AIHA), and the Industrial Medical Association (IMA).

American Industrial Hygiene Association, and the Industrial Medical Association, this guideline is the basis for the occupational noise provisions of the Walsh–Healey Public Contracts Act[9] (see Chapter III, page 74) and the Occupational Safety and Health Act.

When exposure to noise is intermittent rather than continuous, the

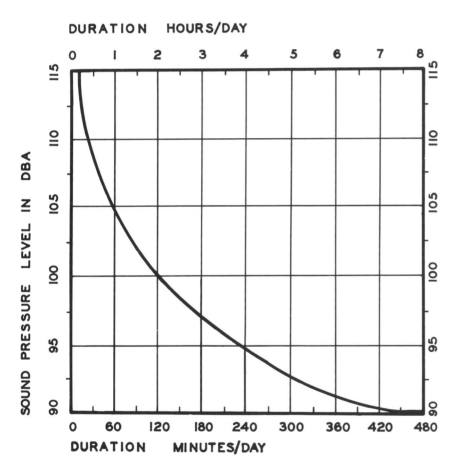

FIGURE 2.4. *Maximum recommended noise exposure in dBA for 8 hours or less. Data from Second Intersociety Committee on Guidelines for Noise Exposure,* Guidelines for Noise Exposure, *October 24, 1969.*

TABLE 2.3.   Acceptable Exposures to Dangerous Noise

Total noise duration per day (24 hours)	Number of times noise occurs per day						
	1	*3*	*7*	*15*	*35*	*75*	*160 up*
8 hrs.	89	89	89	89	89	89	89
6	90	92	95	97	96	94	93
4	91	94	98	101	103	101	99
2	93	98	102	105	108	113	117
1	96	102	106	109	114	125	125 (1¹/₂h)
30 min.	100	105	109	114	125		
15	104	109	115	124			
8	108	114	125		A-weighted		
4	113	125			sound levels		
2	123						

worker can tolerate either greater intensities or the same intensity for longer periods.[10,11] Occasional relief from noise reduces the possibility of noise-induced hearing loss. Botsford has devised a table to correct for intermittent exposure (see Table 2.3).[12] To use the table, when the sound levels are A-weighted, he states:

> select the column headed by the number of times the dangerous noise occurs per day, read down to the average sound level of the noise and locate directly to the left in the first column the total duration of dangerous noise allowed for any 24 hour period. It is permissible to interpolate if necessary.[13]

Duration is also important when evaluating subjective responses to noise. A patterned noise (occurring with consistent regularity) allows the perceiver to generally tolerate the noise level, that is, when it is not intense noise. In contrast with random noise, which has a startling quality, there is less subjective tolerance.

Subjective Measurement of Noise

In addition to the physical noise measurement, several subjective measurement schemes have been developed to evaluate human responses to noise.

PERCEIVED NOISE LEVEL

A commonly used measurement of annoyance is the perceived noise decibel (PNdB) scale that considers both sound intensity and frequency distribution. The PNdB obstensibly compares noises on the basis of their subjective acceptability or their "noisiness."[14] A series of equal "noisiness" contours was developed by Kryter based upon the judgments of equal annoyance for bands of sound one-third octave wide. This scale was initially developed for judging human responses to aircraft overflights, but it has since been applied to other community noises.

Several airports have adopted PNdB thresholds for aircraft operations. The thresholds are noise ceilings that aircraft are required to satisfy. At John F. Kennedy International Airport, planes in takeoff are "theoretically" not to exceed 112 PNdB.[15] The Federal Aviation Agency has adopted a variation of the PNdB (the EPNdB, effective perceived noise) for certifying new generation aircraft, as discussed in Chapter 1.

NOISE NUMBER INDEX To analyze the results of a social survey conducted in the vicinity of London's Heathrow Airport, the Ministry of Aviation devised an index which, it believes, more accurately appraises annoyance responses to aircraft noise than does the PNdB. The Noise and Number Index, as it is called, is a modification of the PNdB since it additionally considers the number of aircraft per day as a key annoyance factor.[16]

This index is defined as:

$$\text{NNI or Noise Number Index} = (\text{Average Peak PNdB}) + 15\ (\log_{10}N) - 80$$

N is the number of aircraft per day or night; the average peak perceived noise level (PNdB) is the highest noise level observed in an overflight; and "80" represents the value subtracted to bring the NNI to 0 or the point of annoyance when using PNdB. To convert PNdB to NNI using the above formula, assuming a PNdB of 105 and an N of 140, you would do the following:

(1) $\text{NNI} = (105\ \text{PNdB}) + 15\ (\log_{10} 140)\quad 80$
(2) $\qquad\qquad\qquad\qquad\quad + 15\ (2.1461)\quad -80$
(3) $\qquad\quad 105 + 32\quad -80$
(4) $\text{NNI} = 57$

As indicated by Figure 2.5, a 57 on the NNI scale of 0-60 would be perceived as more than "Annoying" in terms of "Average Intrusiveness," and in terms of "Average Annoyance" it would be considered above "Moderate" annoyance. Aircraft noise reaches an unreasonable level in the range of 50-60 NNI.

SPEECH INTERFERENCE LEVELS The effect of noise upon speech intelligibility is determined by averaging the readings of three octave bands, 500 Hz, 1000 Hz, and 2000 Hz. The average of the three, expressed in decibels, constitutes

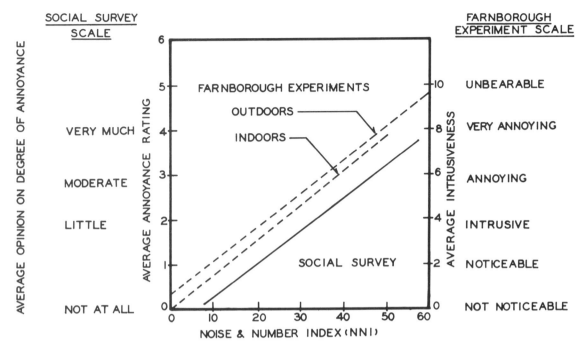

FIGURE 2.5. *Relations between annoyance rating and noise and number index obtained from social survey and Farnborough experiments.* Source: *Wilson Committee on the Problem of Noise,* Noise: Final Report *(London: HMSO, 1963).*

the speech interference level or SIL.[17] There are recommended maximum permissible SIL established for communicating at various distances and in different room environments (see Table 3.3).

TRAFFIC NOISE INDEX

The Traffic Noise Index (TNI) is a method developed to measure annoyance responses to motor vehicle noise, and therefore it is similar in purpose with the NNI, developed for aircraft annoyance. This index was "derived from data representative of traffic noise levels at the facades of buildings varying in distance from the source. It is weighted to take account of variations in traffic flow, and it correlates highly with general dissatisfaction."[18] Traffic noise levels are monitored over a 24-hour period, and the TNI is derived by combining the levels exceeded in dBA 10 percent and 90 percent of the time. Characteristic noise exceeded 10 percent of the time would indicate the extremely noisy vehicles, whereas the 90 percent level represents the average background noise of traffic, the general traffic hum, so to speak.

Specifically the Traffic Noise Index is expressed by formula as:

TNI = 4 (10% level−90% level) + (90% level−30)

Suppose, in recording and analyzing traffic noise, that the 10 percent level observed was 85 dBA and the 90 percent level was 60 dBA. The TNI would be determined as follows:

(1) TNI = 4 (85−60) + (60−30)
(2)   100 + 30
(3) TNI = 130

Griffiths and Langdon have developed a table to compute minimum acceptable distances from a roadway for tolerance of vehicle noise levels (see Figure 2.6).[19] For a TNI of 130 to be accepted by 50 percent of the community residents as Figure 2.6 suggests, there should be a distance of approximately 28 meters or 91 feet between the roadway

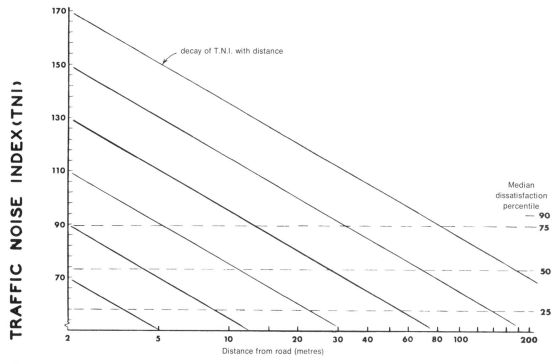

FIGURE 2.6.   *Suggested form of a chart to predict the minimum acceptable distance from the road for various levels of dissatisfaction given one measurement of TNI (external) and distance.* Source: I. D. Griffiths and F. J. Langdon, "Subjective Response to Road Traffic Noise," The Journal of Sound and Vibration *3:10, October, 1969.*

and the residents. To gain a greater percentage of acceptance from the community, let us say 75 percent, would require doubling the distance, from 28 meters to 56 meters when the TNI remains 130.

With such numerous methods for appraising noise and with the complexities of measurement, standards and solutions to noise problems are slow in coming. Equally important, the variety of units for measuring human responses to noise such as PNdB, dBA, and octave bands often produces differences in responses—minor differences, but differences they remain. Recently a way to duplicate certain measuring techniques by using the sound level meter with little loss in accuracy has been described.[20]

The primary contribution of this simplifying measurement is that it eliminates the necessity of taking octave band sound pressure level readings which are a prerequisite for establishing PNdB and SIL index. For example, there is a relatively consistent relationship between PNdB and dBA. A correction factor of 13 can convert a dBA reading obtained on the sound level meter to a PNdB, when the noise source is aircraft. Rather than perform the various steps required to obtain a PNdB level, in the community noise measurements described in subsequent chapters, we measured the A-weighted sound level in dB and converted to PNdB units by adding 13. Consequently, an aircraft reading of 90 dBA was considered to be equivalent to 103 PNdB.

III

The Nuisances and Hazards
of Noise

Noise is part of the environment in which we live. To determine the severity of noise as an environmental concern, some criterion has to be chosen. Health is a logical criterion because it covers all the effects upon the organism, rather than merely the absence of disease. For our purposes, we consider health as a quantitative measure of physical, emotional, and social well-being.[1]

The health problem of noise has two facets. First, noise is a hazard disruptive to the organism's physiology. Noise-induced hearing loss, for example, is principally an organic disability affecting the hearing mechanism of the ear. Second, noise can be an annoyance. It may evoke a feeling of resentment as it intrudes into physical privacy, or into one's thoughts.[2] Annoyance, as a subjective response to noise, is capable of affecting health in terms of emotional and/or social well-being.

Generally, the physiological damage caused by noise is more severe than the psychological problems of annoyance and irritability. But although permanent hearing disability due to noise is of greater concern, noise annoyance should not be overlooked.

The effects of noise on health represent a continuum from hazards to less severe nuisances (see Figure 3.1).[3] Hazards (Stages 1, 2) challenge survival, causing physical injury, while nuisances (Stages 3, 4) reduce human performance or affect human comfort and enjoyment. To rid the environment of noise pollution completely requires eliminating the health effects described in all stages, beginning with Stage 1 ("Insure Survival").

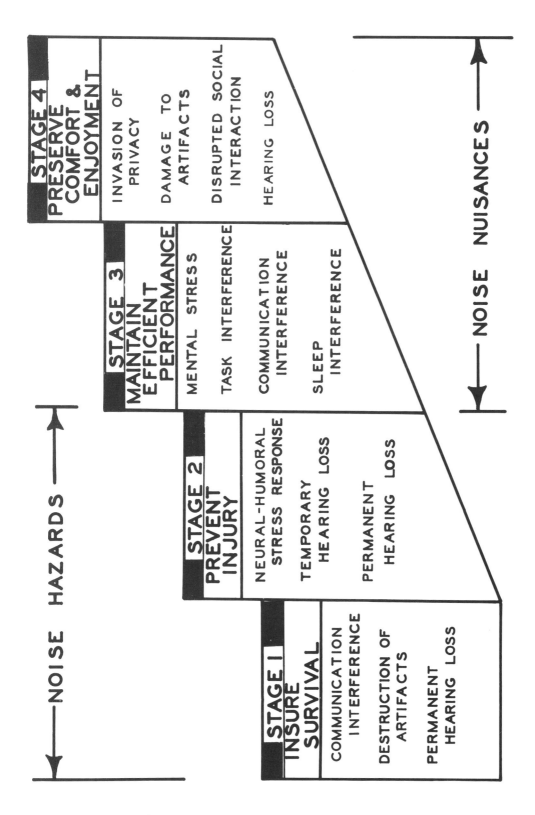

Figure 3.1. *Noise and its implicated health effects*

Protecting human life is a fundamental concern, and occasionally life itself becomes threatened by noise. Fatalities occur because noise interferes with, or masks, important sound messages. Physical survival can also be threatened, although less frequently, when there are sonic disturbances or booms. These sonic overpressures, generated most often by supersonic aircraft, have been known to collapse homes and destroy historically important landscapes. Besides producing such physical danger, noise is capable of affecting economic well-being by reducing earning power. Severe noise-induced hearing loss may affect the ability of an individual to perform the job for which he is trained. Temporary and possibly permanent unemployment can result.

Stage 1: Insure Survival

Noise that interferes with communication can be hazardous, particularly when a message intended to alert a person to danger is masked by noise. Two people were killed and several injured the day Senator Robert F. Kennedy's southbound funeral train passed through Elizabeth, New Jersey.[4] They were unable to hear the warning horn of a train approaching from the opposite direction, and therefore did not get out of its way. It was later determined that secret service and news media helicopters had completely obliterated the signal warning of this train's approach.

COMMUNICATION INTERFERENCE

Many vehicular accidents occur because the driver is unable to hear an emergency vehicle signal. The siren's warning is masked by the noises heard inside a car or truck. The situation has reached a point where, in the opinion of one state official, "more lives are lost because of speeding fire apparatus, police cars, and ambulances than are saved by this practice."[5] Up to the present time, trucks posed the largest problem, since interior noise in trucks was greater than that registered within automobiles. When an emergency vehicle is in the process of passing a truck, the driver can detect the siren's warning for only a very short period, three seconds or less (see Figures 3.2 and 3.3).[6] During the remaining time, truck noise drowns out the signal, and sound messages go by undetected.

Noise that masks hearing is but one problem affecting driver performance. Noise can also be distractive to the individual who is driving. There are wind, engine, and tire noises. Air conditioners and stereophonic tape/radio units, as well as other automobile accessories, create distractions that reduce driver concentration. The driver becomes less aware of his immediate environment. Driver attentiveness to visual and auditory cues probably lessens under these distracting conditions. In fact, some traffic safety authorities believe that auditory distraction is so severe that drivers with limited or no hearing are better risks than those with normal hearing. Some contend their accident rate is lower on a comparative basis.[7] Ontario Department of

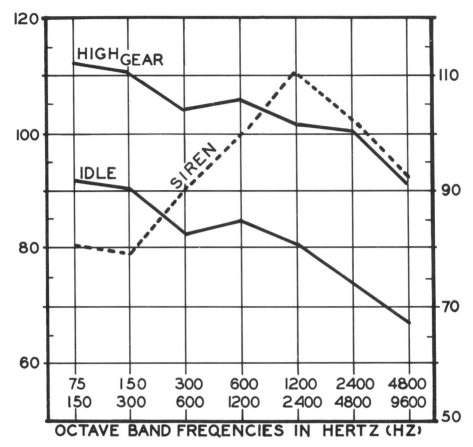

FIGURE 3.2.   *Octave band analysis of truck noise and ambulance siren.* Source:
*Paul S. Veneklasen, "Summary: Community Noise Control," in D. W.
Ward and J. Fricke, eds.,* Noise as a Public Health Hazard *(Washing-
ton, DC: American Speech and Hearing Association, 1969).*

Transport officials report deaf students are the better drivers in the
province. Of the few studies conducted, deaf drivers perform at least
as well as those with hearing.

The military recognize that aircraft noise masks essential communications. Helicopter noise is particularly hazardous to pilots in combat
situations, primarily where radio communications involve the lives
of the crew and troops they are supporting. Although no statistics are
available, one authority believes that the intense noise levels inside
helicopters, which disrupt communications, have been responsible
for pilot misjudgments resulting in the loss of human life.[8]

Noises have masked warnings and compromised safety in other

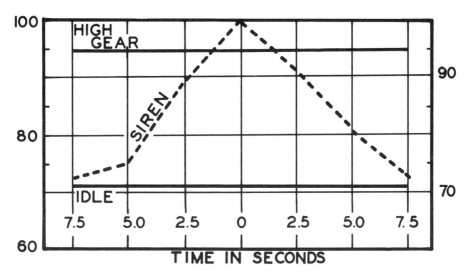

FIGURE 3.3. *Sound level of siren inside truck cab. Data from Paul Veneklasen, "Summary: Community Noise Control," in D. W. Ward and J. Fricke, eds.,* Noise as a Public Health Hazard *(Washington, DC: American Speech and Hearing Association, 1969).*

activities besides transportation. Heavy industrial operations offer the type of environment where communication interference is a problem. Again, little documentation exists, at least for public dissemination. Although records denoting industrial accidents of this type are unavailable, in all likelihood such accidents do occur with some frequency. In the occupational environment Freed reports that deaf people, who are not influenced by surrounding noise as well as other stimuli, have fewer accidents.[9]

DESTRUCTION OF PHYSICAL OBJECTS

The survival of physical objects, which consist of both natural and man-made environmental features (natural land characteristics, dwellings, transportation systems, etc.), is being increasingly challenged by our ever-expanding technology of which noise is a by-product. Destructive overpressures are being generated by supersonic aircraft flying over land areas. Although most damage to date is of a minor nature (see "Damage to Physical Objects," pages 87–89), some of it is irreparable.

In two national parks, 80 tons of rock in the form of prehistoric cliff dwellings and natural sandstone formations succumbed to the boom and were permanently destroyed. Investigating these incidents, the National Park Service reported to former Secretary of the Interior Stewart Udall:

These important landmarks of our heritage have been saved through bitter controversies from the wanton hand of the earth-bound despoiler. The cliff-dwellings have survived the ravages of centuries because of their sheltered location in caves and overhangs. Now these natural formations and the remains of man's habitation upon the land are threatened by invisible shock waves of speeding aircraft. Surely we would not stay the hand of our technological process. At the same time, however, we must not allow inheritance to be defaced or destroyed through thoughtless or careless acts of individuals.[10]

Older structures appear to be most easily affected. Major structural damage is possible, although the chances of its happening remain slight. However, in the fall of 1967 a French farmhouse totally collapsed, killing three occupants.[11] Seconds after the survivors heard a loud clap attributed to a sonic boom, the rafters gave way, sending down eight tons of barley stored in the attic. Although in no case has evidence been entirely conclusive, sonic booms have been implicated in the death of eleven French citizens since 1963.[12]

If and when commercial supersonic flights are authorized over land, the probability of extensive damage, particularly to delicate geographical features, historic structures, and older dwellings in need of repair will be increased. As long as the noise problem persists, it will be a challenge to the survival of society's artifacts.

PERMANENT HEARING LOSS

It seems probable that a person suffering from occupational hearing loss would run the risk of future job insecurity. If he should choose to change jobs, his hearing loss might not be as well accepted in a different position. Obviously the worker is barred from any employment requiring hearing acuity. Furthermore, it is known that some employers discriminate against the hard of hearing because of alleged higher insurance costs.[13] Not only does the employer use the insurance rate argument to discriminate, but so does the company physician.

. . . a placement agent advised the American Hearing Society that the employer in a certain factory was willing and ready to hire a hard-of-hearing person, but stated that the examining physician would not pass the worker for fear that if an accident occurred, he, the physician, would be blamed. Such practices are unfair to physically handicapped persons, and defeat the very purposes of the law. A workman's compensation premium rates are based upon hazards inherent in the industry and the experience record in the payment of benefits for work injuries sustained over specific periods of time.[14]

This problem of employment security becomes greater with advancing age. Elderly workers who have incurred hearing losses are doubly at a disadvantage with hearing loss and age both against them.

Economic survival may be a more immediate problem to them than to younger workers with a loss in hearing.

Statistics indicate that, whatever the age, the deaf and hard of hearing commonly hold the lower paying, less than skilled jobs, and this despite the fact that they possess the same intelligence as hearing people.[15] Today four out of every five deaf persons are performing manual labor work as compared to one half of the general population.[16] With the trend toward white collar skilled employment opportunities, those suffering a major loss of hearing are competing for a decreasing number of jobs that proportionately pay less and less.

Noise is capable of acting as a physiological stress, by contributing to the onset of injury or disease. The term "stress" implies a change in conditions affecting an organism which requires effort to maintain essential functions at a desired level. Because a load (such as noise) is placed upon the organism, the organism must modify its behavior to continue operating. This involves adaptation or a compensatory response which enables the organism to function despite the load. The compensatory or adaptive response to a noise stress can cause significant physiological changes that on occasion lead to human dysfunctioning. Adaptive techniques, though necessary to maintain body functioning, are often warnings that more serious responses may ensue.[17] A ringing in the ears, for example, represents a physiological response to intense noise, warning the individual that continued exposure could permanently affect his hearing threshold.

Noise is a stress which is often intense, repetitive, and/or of long duration. The body accordingly responds with a variety of hormonal and neurological mechanisms. It is believed that some of these responses—when prolonged, repeated, or of intense levels—may contribute to the development of progressive psychosomatic disease. Noise may be a factor among such diseases as peptic ulcer and essential hypertension. Although the validity of the evidence must be carefully reviewed, Michael's summary of findings indicates that colitis, high blood pressure, migraine headaches, and nervous disorders are health effects which can be associated with noise.[18] Noise as a stress may also worsen or perpetuate a disease after it has developed. One hospital survey, prepared for the U.S. Public Health Service, concludes that hospitalization among convalescing patients is possibly prolonged in

Stage 2: Prevent Injury and Disease

NEURAL-HUMORAL STRESS RESPONSE*

---

*There are various nonauditory somatic responses to noise reported in the literature. The proposed hypotheses to explain these responses to a noise stimulus include neurological, hormonal, and neurohumoral mechanisms. The term neural-humoral is used to describe the above responses.

noisy environments.[19] Intensive care patients are most affected by intrusive noise.

There is general consensus in the literature that there are physiological responses associated with noise. Particularly noted are the responses in man's vegetative or autonomic system. Broadbent, among others, has summarized these responses. Unexpected noise may produce a rise in blood pressure, an increase in pressure within the head, greater heart and respiratory rates, and sharp muscular contraction.[20] Digestive functioning, too, can be disrupted. Short noise exposures may well decrease the contractions conveying food through the body and regulating the flow of saliva and gastric juices.[21] Peripheral constriction of the blood vessels persists as a normal response to repeated moderate noises, according to several sources, with little adaptation occurring.

Key questions then are not only what physiological responses transpire under conditions of noise, but also whether the changes are temporary or prolonged. Do adaptive techniques precipitated by noise create pathological or irreversible results which consequently impair our health? What noise levels are necessary for such responses to occur?

First let us examine some of the autonomic responses that generally are attributable to a noise source, disregarding for now the questions of degree of response and specific noise intensity. Central to this discussion is the important assumption that *until someone proves that these changes are negligible, we must assume that the physiological effects of noise on health are hazardous.*

Effect on Cardio-Vascular System (Circulation)

It is generally thought that the circulatory system is more significantly affected by noise than any of the other life systems. Peripheral vasoconstriction, which narrows the terminal, very muscular arteries (arterioles), is a well-documented circulatory response to noise.[22] The finger pulse amplitude test (digital plethysmography) indicates the extent of peripheral blood constriction. A decrease in circulation to the extremeties or a comparable increase in arterial blood pressure results. Upon exposure to noise, arteriole contraction appears to be more sustained than muscular tension or increased heart or respiratory rates.[23] Present evidence indicates only minor adaptation with moderate, often repeated noise. Generally, as intensity increases adaptation decreases. If vasoconstriction is sustained, the decreased blood volume reduces the available oxygen vital to human activity. Under this condition, noise-induced fatigue may ensue.

When tension accompanies a lower blood volume, further health impairment follows, according to a Council of Europe report. A chain reaction may influence glandular activity and metabolic functioning.[24]

A Soviet study reports that in high intensity noise, there is a loss of circulation to the brain and hence a decrease in the oxygen level to cerebral tissue.[25] Unfortunately, no other studies of this type have been undertaken to substantiate these findings. Anything beyond the fact that blood volume decreases while arterial pressure increases, under the influence of noise, is at present considered speculation or conjecture.

The United Nations World Health Organization has cited Soviet research that found a high number of gastric complaints among groups of people subject to prolonged intense noise.[26] This writer is unaware of any United States research in this area. According to research conducted in England, noise does interfere with digestive functioning particularly when the noise is sudden and unexpected. Such noise produces a marked sympathetic nervous system response, resulting in a decrease in bowel activity, saliva flow, and digestive juices.[27] Under repeated exposure, digestive upset may continue, though to a lesser extent. Lehman, in more general terms, notes that the secretion and composition of gastric juices is altered.[28]

Effect on Digestive System

Additional physiological responses are reported in the literature. Heartbeat is often slowed. Respiratory functions such as breathing may quicken, and a higher galvanic skin response usually appears, especially when the noise is unexpected, such as a sonic boom. Although the reaction appears to be short, the muscular system undergoes a change. Again, primarily with unexpected noise, muscles tend to contract. But adaptation occurs promptly. Adrenalin also speeds through the body and chemically stimulates or triggers physiological changes.

Effect on Other Systems

Recognizing that alterations in the complex physiological system do take place, it is important to know if any of these responses to noise become irreversible, leading to a pathological health effect. In discussing potential pathological or irreversible health effects, one must consider adaptation as well as noise intensity. Adaptation, as used here, is the ability of a human to adjust to a given noise stress without its adversely affecting his state of health. Apparently the ability to adapt differs for persons sleeping and for those awake. The little work done to date suggests that there is a lower threshold level during periods of sleep. In other words, noise is more physiologically disruptive while one is sleeping.[29,30]

Until recently, the best evidence concluded that although autonomic responses occurred upon noise exposure, there was little likelihood that the noise would impair normal functioning. Second, the only noticeable responses to noise began at 90 decibels or above. Very revealing work by Jansen indicates that neither conclusion is valid.[31] Vasoconstriction and pupil dilatation begin to appear at 70 decibels, and only up to 95 decibels do no pathological side effects necessarily appear. However, there is the opinion that hazardous side effects are very probable above the 95 decibel level, and depending upon the duration of exposure. We must keep in mind that the assumptions made in setting standards involve only the average "normal population." For those already suffering from some additional psychic or somatic stress, Jansen contends, in all likelihood a lower level (below 95 dB) would effect a pathological response.[32]

Noise duration has always been a factor in the ability to adapt. Broadbent, among others, concluded that in noise below 120 dB people adapted to long exposures, with no noticeable physiological changes.[33,34] The results of a three-year study are in contradiction. Noise-induced vasoconstrictive effects appeared with repeated noise bursts lasting from 300 milliseconds up to 60 minutes.[35] A selected population exposed to 30 to 60 minutes of white noise* (90 dB) daily for three years developed no adaptation. Their physiological responses remained constant while they were being tested.

More evidence is necessary to determine the total effect of noise on man's physiological system, including its role in either aggravating or precipitating illness. Certainly the answers must be sought through studies with a set population monitored over an extensive time period. Repeated observations will help to answer the question of how noise duration is related to adaptation.

The community noise levels experienced by an urban population definitely fall into the range where they produce physiological responses and possible health impairment. Jansen strongly advocates the use of risk criteria as a health guideline. Generally he suggests that the maximum wide band noise level (common in community environments) should not exceed 88 decibels, regardless of duration.[36] Refinements may come, but at present we would say that this decibel level must not be exceeded if we are to remove the hazard potential of Stage 2.

HEARING LOSS

Today it is well recognized that hearing loss is statistically the major hazard of a noisy environment. Within working environments, loss of hearing is possibly the most common occupational hazard. Glorig suggests that approximately 6 million workers suffer from impaired hearing due to continuous high intensity noise throughout their occupational lives.[37]

Unfortunately there appear to be no comprehensive data regarding

*Where noise energy is uniform over a wide frequency range.

the extent and degree of hearing loss among the total population. Surveys to date have been conducted only on specific age groups.[38] Numerous United States estimates based on limited samples do exist, however. For instance, Braun has estimated that 8 to 10 million Americans need hearing aids.[39]

A U.S. Public Health Service study indicates that in 1963 there were 6.5 million persons suffering from hearing loss in one or both ears.[40] The Save Your Hearing Foundation believes that hearing disability now involves 15 to 40 million persons.[41] Although noise is just one cause of hearing loss and therefore cannot be accountable for all impairment, there are several reasons for being concerned with it from an environmental health standpoint.

First of all, an increasing proportion of the population is being exposed to noise in the environment. This increase is due not only to rising numbers in the work force. Equally important is the work environment itself.

A significant amount of evidence has implicated the industrial environment as the primary cause of noise-induced hearing loss. Studies comparing industrially exposed with nonindustrially exposed populations show a hearing loss difference of between 10 and 30 percent.[42] On the average, the chances of suffering hearing loss while employed in industry are as much as one third greater than they are in a nonindustrial employment. The U.S. Public Health Service estimates that there may be 16 million persons exposed to noise levels that are hazardous to hearing.[43]

No longer can we attribute hearing loss solely to the work place. The entire community must be considered suspect. Although there is not complete agreement on this position, Dougherty and Welsh have remarked:

> Recent population studies have suggested that hearing loss formerly thought to be a hazard of aviators and boilermakers occurs with age after lifetime exposure to noise at a community level.[44]

Community noise levels often exceed industrial working noises. However, there are difficulties in drawing conclusions about how much community noises affect hearing. Many variables must be considered (noise intensity, frequency distribution, and length of exposure). Furthermore, no consensus of opinion exists on what are physiologically tolerable noise levels. As evidence accumulates, though, damage risk criteria, proposed since 1940, are becoming progressively lower.[45]

The American Academy of Ophthalmology and Otolaryngology considers that at this time knowledge of the relationship of noise-

TABLE 3.1.   Recommended Frequency Range for Instituting Hearing
Conservation Measures

Organization	Decibel level	Frequency range
American Academy of Ophthalmology & Otolaryngology	85 dB	300–1200 Hz
British Medical Association[46]	85 dB	250–4000 Hz
U.S. Air Force[47]	85 dB	300–4800 Hz
U.S. Army	85 dB	300–9600 Hz

exposure to hearing loss is much too limited for proposing "safe" amounts of noise exposure. They do suggest, however, that if habitual exposure to a continuous steady-state noise at 85 decibels occurs, the hearing of personnel exposed should be evaluated.[48] There is general agreement that 85 dB is the threshold level for hearing damage and the wearing of hearing-protective devices. The major difference of opinion arises over the frequency range believed hazardous (see Table 3.1). The U.S. Army guideline (TB Med 251) recommends a hearing conservation program when the noise level exceeds 85 dB in any octave band between 300 and 9,600 Hertz.[49] This recommendation is the most restrictive of the four in Table 3.1.

Under the provisions of the Walsh-Healey Public Contracts Act and the Occupational Safety and Health Act occupational noise exposure levels have been set.[50] The OSHA affects businesses in interstate commerce (involving 57 million workers). Protection against the effects of noise must be provided when the overall noise level exceeds 90 dBA over an eight-hour period. This corresponds to octave band levels of: 86 dB at 4,000 Hz, 87 dB at 2,000 Hz, 88 dB at 8,000 Hz, 89 dB at 1,000 Hz, and 92 dB at 500 Hz (see Figure 3.4). The entire worker population under such contracts will not be protected under the present 90 dBA Walsh-Healey noise requirement. Between 10 and 15 percent of these workers will remain unprotected and should expect to receive a hearing impairment in the speech frequencies (500 Hz, 1,000 Hz, and 2,000 Hz) after thirty years of continuous eight-hour exposure.

There is evidence that even standards set at the 85 dBA level are too high to protect human hearing. Several authorities suspect functional hearing loss at 80 dBA.[51,52] Below 80 dBA, chances of experiencing any permanent loss of hearing in the speech frequencies are slight. In analyzing 6,835 audiograms of workers, Baughn observed no incidence of loss at 80 dBA.[53]

But what about hearing loss from nonoccupational noise exposure, noise occurring during the time beyond the normal eight-hour work day? Cohen *et al.* has proposed noise limits for nonoccupational conditions (see Table 3.2).[54] These are based on the intent to preserve

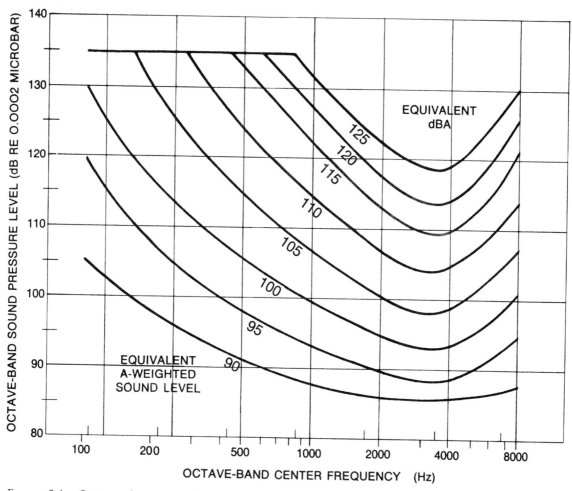

FIGURE 3.4. *Contours for determining equivalent A-weighted sound level.* Source: Federal Register, *May 10, 1969.*

hearing in all audiometric frequencies between 500 Hz and 6,000 Hz, rather than in just the speech frequencies (500–2,000 Hz). Conditions become potentially unsafe when the off-job noise exposure exceeds 70 dBA during sixteen or more hours a day.

There are reports that in the U.S.S.R. personal protection is required for noise levels even below 80 dB. In Russia, noise levels in industrial work environments are not allowed to exceed 70 dB unless hearing protection is provided to the worker.[55] It is not known if such a requirement is enforced.

Hearing affected by noise represents a sensory-neural loss. When

TABLE 3.2.   Noise Limits for Nonoccupational Noise Exposures

Limiting daily exposure times for nonoccupational noise conditions	Sound level in dBA
<2 minutes	115
<4 minutes	110
<8 minutes	105
15 minutes	100
1/2 hour	95
1 hour	90
2 hours	85
4 hours	80
8 hours	75
16–24 hours	70

Source: Alexander Cohen *et al.,* "Sociocusis—Hearing Loss from Non-Occupational Noise Exposure," *Sound and Vibration,* 4:11, November, 1970.

the noise is of sufficient intensity and duration, a permanent hearing loss results. This loss is referred to as a noise-induced permanent threshold shift (NIPTS or PTS). A PTS causes irreversible damage to the inner ear.[56] There are no known ameliorative agents able either to inhibit or to cure this type of hearing loss.[57]

The site of noise-induced hearing loss is the Organ of Corti, located within the cochlea of the inner ear. Damage to this organ occurs in the following manner:

> Sound induced motion of the fluid in the cochlea induces shearing and bending movements of the hair cells in the Organ of Corti, which, in turn, result in electrical stimuli transmitted by the auditory nerve. Prolonged and excessive noise eventually produces deterioration and, finally, destruction of hair cells, and thus disrupts the sound transmission mechanism.[58]

Not every hearing loss or shift in hearing threshold is permanent. A temporary threshold shift (TTS) occurs usually when there is a short exposure to noise. TTS in hearing acuity is any loss from which the ear recovers. Normally, recovery takes place within twenty-four hours.[59] In contrast, with more severe and repeated exposure, recovery may not be complete, and a permanent threshold shift (PTS) remains. It seems likely that some relationship exists between PTS and TTS in the onset of hearing loss. But a test devised for predicting the magnitude of a PTS based upon TTS has been challenged.[60] Ward and Nelson have found that among chinchillas there was no significant statistical correlation between the two.[61] To date no study has directly examined noise-induced PTS and TTS among humans.

A threshold shift in hearing can be suspected from the characteristic loss of speech understanding. At that point, however, irreversible higher frequency loss has been incurred.

There are other dangers connected with a PTS besides its irreversibility and insidious qualities. Unlike physiological reactions to many other stresses, the hearing mechanism doesn't "toughen" as an individual continues to be exposed to noise. No callouses develop. In a consistently noisy environment, auditory fatigue (threshold shift) continues to increase with the passing of time. Little if any adaptation can be observed; consequently the long-term effect is cumulative damage.

At Stage 3 the environment is no longer considered hazardous. Noise associated with this stage and Stage 4 has only a nuisance effect upon health. In Stage 3 it is capable of interfering with work, communication, sleep, and normal behavior. The removal of the nuisance potential of noise from the environment provides an opportunity for the individual to perform more efficiently.

Stage 3: Maintain Efficient Performance

While noise may not be intense enough to damage hearing, its interference with communication can be disturbing:

COMMUNICATION INTERFERENCE

> . . . from a health point of view the auditory damaging effect of noise is, of course, deplorable. The temporary interference of noise with the proper functioning of the auditory system during man's daily work and social activities are, from a practical point of view, perhaps more important and cause more suffering.[62]

The reason is that communication gives us sound messages essential to operating efficiently.

Initially, hearing loss occurs in the upper frequencies. As a rule this loss is for frequencies above 3,000 Hz. However, hearing frequencies above 3,000 Hz are essential for clearly understanding speech. In the early stage of deafness, hearing loss is damaging but usually goes unnoticed. An individual may experience as much as 40 percent bilateral hearing impairment without being subjectively aware of it.[63] Noise is an insidious threat to hearing acuity. Medical examination remains the most accurate way of detecting such a loss.

For ease in listening, over 90 percent of the words spoken should be correctly heard. Listening starts to be fatiguing when the percentage heard is lower.[64] Most speech sounds register below 3,000 Hz; therefore hearing loss in those frequencies must be avoided. However, the primary spectral distribution of speech sounds falls between 200 and 6,400 Hz[65] (see Figure 3.5). To hear speech completely, Beranek contends that all sounds in the frequency range from 200 to 6,000 Hz are essential.[66] Webster, on the other hand, recommends that hearing within the range of 100 to 7,000 Hz is essential.[67]

Several factors affect the choice of an upper frequency level. One is

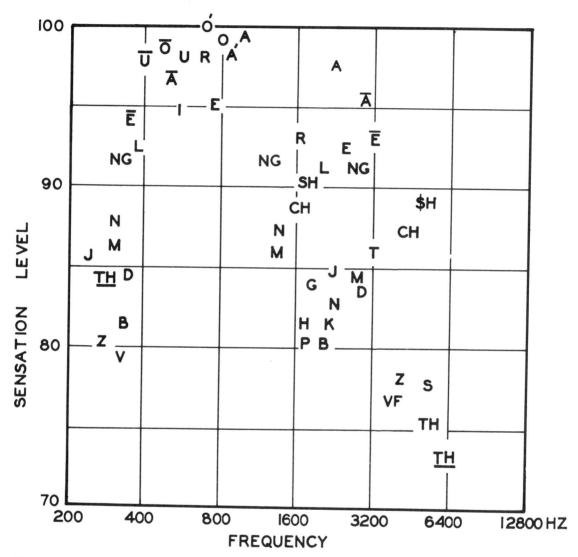

FIGURE 3.5.   *Combined characteristics of the fundamental sounds of speech.* Source: *H. Fletcher,* Speech and Hearing in Communication *(New York: Van Nostrand, 1953). Copyright © 1953 by Litton Educational Publishing Inc. Reprinted by permission of Van Nostrand Reinhold Company.*

the ambient or background noise level. Traditionally hearing tests are conducted in quiet laboratory conditions, and under these conditions the discrimination of speech requires hearing frequencies no higher than 3,000 Hz. However, in a typical city environment where ambient noise is greater, speech discrimination is more strongly affected; it

their speed decreased, and they obtained lower test scores. In other research, noise has been found to stimulate, or at least not to hinder, students with higher than average IQ's. Those of average or below average IQ were unable to perform as well in a noisy environment as in a quiet one.[94] On the other hand, Corso *et al.* found students to be unaffected by noise.[95,96] Many controls are necessary in these experiments before consistent results will be reported.

Until now, we have discussed performance in moderate but not excessive noise (90–110 dB). At these levels, mental performance is generally affected in some manner, but lower-order tasks requiring physical or mechanical faculties only are less affected.

But when noise becomes extremely intense, or above 120 decibels, both mental and physical performance are severely curtailed. Unless protective hearing devices are worn, there is noticeable pain and damage to the ear. There also may be unpleasant body sensations that include internal head vibrations, air movement within the nose, loss of equilibrium, and vision disturbance.[97]

Although there is less evidence, some are of the opinion that the threshold for performance impairment lies below 90 decibels.

> Many, perhaps most, ordinary people would say that their depth of concentration, business efficiency, output, etc. are affected by noise well below the 90 dB level. This common sense opinion must be respected even if it is not supported by such carefully controlled experiments as have so far been made.[98]

The World Health Organization has given some support to this position. They estimate that office noise (which rarely exceeds 90 dB) causes inefficiency amounting to $4,000,000 every workday. These levels found in offices throughout the world contribute to misunderstanding of auditory communication, fatigue, absentmindedness, and mental strain.[99]

A final factor in considering noise and task interference is the definition of performance. If performance is defined simply as the ability to complete a task while exposed to noise, without consideration for associated stress, the total effect of noise will not be fully evaluated. In nearly all the experiments performed to date, emphasis has been on evaluating whether or not the primary task has been achieved under certain noise conditions. The question of compensatory responses on the part of a subject in order to complete a task is largely overlooked. A very revealing study by Lehman points up this significant limitation of performance studies. In examining a group of German workers, he found that "those workers in noisy jobs tend to be more quarrelsome at work and away from it than those doing equivalent jobs but who are not subject to similar noise stresses."[100] It is obvious that we must look beyond a single environment or a particular goal of work to assess fully the role of noise in affecting human performance.

MENTAL STRESS   The question of mental stress and the role of noise is highly controversial. Unfortunately there is little substantive literature. Popularized journalistic accounts frequently suggest noise as a causal factor in creating mental health problems. Sometimes these stories have merit. In New York noise was cited as a major reason for the shooting of a young boy:

> . . . one evening last spring, four boys were at play, shouting and racing in and out of an apartment building. Suddenly, from a second-floor window, came the crack of a pistol. One of the boys sprawled dead on the pavement. The victim happened to be Roy Innis Jr., thirteen, son of a prominent Negro leader, but there was no political implication in the tragedy. The killer, also a Negro, confessed to police that he was a nightworker who had lost control of himself because the noise from the boys prevented him from sleeping.[101]

Many have responded to this type of reporting by discounting any possibility that some relationship may exist; but this position also seems unrealistic since "categorical statements about the lack of mental dysfunction caused by exposure to noise are at the moment premature."[102] No definitive conclusions either way can presently be made. More than likely noise is just one environmental factor influencing the population's mental health.

Under certain circumstances the likelihood of mental health impairment due to noise may be greater. One situation, according to a Council of Europe report, is that in which the individual has a disposition to nervousness.[103] Another is noise aggravating an already existing neurosis. In the case of a predisposition to mental stress, noise tends to aggravate the condition.

This potentially "noise sensitive" population may be large. A specific number is not known, since the prevalence rate has varied among the 35 community mental health studies reviewed by Dunham.[104] The psychiatric profession estimates that approximately 10 percent of our population is in need of or could benefit from mental health care. With this population, health problems may well appear below 95 dB.[105]

There is little evidence to suggest that the general population may suffer behavioral disorders as a result of noise. Few comprehensive investigations of the nonauditory effects of noise have been undertaken. In 1958 a psychiatric investigation among members of two U.S. Navy carriers found no evidence of mental health upset attributable to noise.[106] However, one psychologist observed that the fact that the carrier crew was "selected, well trained and hardened for combat operations" limits the "generalizability of this finding."[107] A community noise study in London found a high mental hospital admission rate in a neighborhood with the greatest amount of aircraft noise.*

Mild behavioral reactions do occur. As Lieber suggests, "Any person who is exposed to a high noise level to which he is not accustomed will at first only suffer a mild discomfort, but after a time he will be subjected to changes of mood. Emotional responses may become more extreme, and it is not uncommon to fly off the handle at the slightest provocation."[108] Michalova and Hrubes found that industrial workers complained neurotically in constant noise above 110 dB.[109] Where there are noisy activities in the general community environment, similar problems arise. Habitual complaints occur in airport communities with a maximum threshold level between 95 and 105 PNdB, or 82 and 92 dBA.[110]

In summary, current informed opinion more than any quantitative evidence indicates that noise may well be a factor in mental stress. More than likely, noise is aggravating rather than precipitating behavioral disorders. The population most susceptible to this influence are those already having some mental health problem, in other words an already problemed population. Although individual susceptibility is the main factor, other factors to be considered are noise intensity, duration, frequency distribution, and pattern of occurrence.

At this stage, noise affects more people than at any other. Problems arising here are an early warning that more severe health effects may appear if noise is not abated.

Stage 4: Preserve Comfort and Enjoyment

People experience the intrusive quality of noise when it is an integral part of their daily activity pattern. Especially among the urban population—whether at work, play, or rest, in the home, on the job, or in between—the chances of finding refuge from noise exposure are rapidly disappearing. With greater and greater regularity, noise is becoming a 24-hour-a-day problem infringing upon the comfort and enjoyment of living. According to one estimate, over 150 million persons, just in their home environment, are subjected daily to annoying noise levels.[111]

What do we mean by comfort and enjoyment? A simple definition is difficult, for they mean many things to many people. However, comfort and enjoyment seek to insure the right to privacy and freedom from intrusion. If these rights are not adequately protected against noise, annoyance results.

---

*Abey-Wickrama et al., "Mental Health Admissions and Aircraft Noise," *The Lancet,* Dec. 19, 1969, pp. 1275-78.

Annoyance constitutes the largest environmental health problem associated with community noise exposure. Some assert that annoyance is a biological protective mechanism, like the discomforts of fatigue, hunger, or cold,which impels the organism to avoid noise as do the other signals of disturbance.[112]

Given the same noise intensity and source, human annoyance will vary considerably. One man's music may be another man's noise. This is so because people associate the hearing of noise with a subjective network of attitudes, expanding the possibility of responses. For example, a manufacturing executive listens to the noise associated with production as a positive sign of company growth and profit. He will not perceive it as "noise." In contrast, a person living next to his cacophonous operation may become highly annoyed, and provoked enough to take action.

Generally speaking, the variation in human responses becomes smaller with higher noise levels. That is, a consensus of opinion grows, regardless of social differences, as noise increases. Furthermore, the level of annoyance does not appear to diminish with time. The London survey found that, with time, objection to community noise grew rather than declined.[113]

There are several reasons why annoyance should be examined carefully as an index of comfort and enjoyment.

TERRITORIALITY: INVASION OF PRIVACY

Unlike light and smell, noise possesses a penetrative ability that respects few boundaries. Visual privacy is much easier to achieve than auditory privacy. Physiologically, there is no way to shut the perception of hearing off completely. The sense of hearing is a twenty-four-hour, 360-degree sense. It is a simple matter to close one's eyelids. But man has no corresponding "ear lids." Man's auditory privacy is thus constantly challenged and invaded by unwanted sounds.

The right to privacy has historically been considered a fundamental legal right—a man's home is his castle. More recently, privacy in the sense of territoriality has been suggested as a basic human need. Certain scientists believe that both lower animals and man instinctively use certain prescribed spatial areas called territories for interacting.[114] Different levels of territorial space are used for different personal and social activities. To insure normal human functioning, certain territorial boundaries must be maintained.[115,116]

Extreme territorial disorganization is now becoming recognized as an indicator of mental disorder among certain schizophrenic types, patients who object violently to being approached too closely.[117] To a normal person, this area would be social interactional space. The treatment is considered successful when such patients are able to tolerate normal interactional distances.[118]

When noise invades some people's personal territory, it is disconcerting and heightens their anxiety. There may well be stages of territorial awareness, associated with both vision and sound, that affect the response pattern to intrusion. Insuring auditory privacy is a therapeutic device, which in all probability helps allow the individual to cope with less controllable noise and other environmental stresses that are experienced particularly in social space.

Noises generated outside and inside the home intrude upon the comforting boundary of acoustic privacy. The notion that the home represents the last refuge from noise is no longer valid. Several investigators conclude that even interior residential noise levels often approach and exceed intensities found within industries and in communities outside. A trend may be in the making that could begin to re-establish territorial limits to invading noise. Several southern California home builders are incorporating a "retreat room" into their building design: a room adjoining the master bedroom but isolated from the rest of the house and, hopefully, from noise.[119] In hospital planning, the elimination of unwanted sounds is becoming recognized as an integral part of patient privacy.[120]

An early sign of territorial invasion is annoyance, which in itself is discomforting. Such a signal demands a response. It seems likely that territorial security is important to our normal state of health, and noise, like other environmental problems, challenges this needed security.

DAMAGE TO PHYSICAL OBJECTS

Any natural or man-made features in the environment (land features, dwellings, transportation-communication systems, etc.) have become increasingly vulnerable to an ever-expanding technology of which noise is a by-product.

A significant amount of litigation has been generated by noise, principally as it affects residential property. Several legal concepts are significant in noise litigation including nuisance, constitutional damaging, trespass, injunctive relief, and the taking of property.[121] The latter is the most common. Here noise represents a taking or pre-emption of property without just compensation. Normally associated with this taking is an alleged decrease in property value. Just compensation usually sought by the plaintiff is to retrieve that amount of loss due to noise.

Many sources of noise have been the cause of lawsuits, ranging from barking dogs, transistor radios, and industrial plants to vehicular traffic and aircraft. Transportation noises cause the greatest number of court cases. The bulk of these concern aircraft noise, and the number is rising rapidly. In 1962 nineteen of our nation's airports were involved in 2,000 lawsuits. Damages sought by homeowners then amounted to about $14.5 million.[122] By 1968, the dollar value had climbed to over

$200 million, with nearly twice the number of airports as defendants.[123] Claims for alledged noise damage are now estimated to be approximately $4 billion.

If a case is decided in favor of the plaintiff, the amount awarded is a percentage of the property's value. In most instances this is well under half of a property's fair market value. Harr notes that "in Oregon, where a liberal damage rule prevails, the Port of Seattle settled with 44 of 196 plaintiffs for $150,000, an average of about $3,000 per affected property."[124] The amount sought can be much higher; for example, ten families sued the city of Los Angeles and their International Airport for $400,000, or $40,000 per affected property.[125] However, as a rule, both the size of the suit and the settlement are appreciably smaller.

Other buildings, besides homes, are affected by noise and the tenants have sought relief. Public buildings and churches can be added to the list. In at least two Florida communities, school districts filed suits against the airport authorities because the noise disrupted the process of education.[126] The city of Boston has a major suit pending against the airlines and the airport operators. Mayor Kevin White is seeking $10.2 million in damages from 19 airlines and from the Massachusetts Port Authority, operators of Logan International Airport. Of this total $4.5 million is sought for soundproofing fifteen east Boston schools. The remaining amount is for the taking of air rights over city property without just compensation, depreciation in the value of both the school buildings and land, limiting the further use of this land, and rendering the schools unfit for educational purposes.[127] Aircraft noise at Los Angeles International Airport (LAX) has generated approximately $3 billion dollars' worth of claims. This total includes a $95.8 million suit by the Inglewood Board of Education against LAX. Airport noise suits are not strictly confined to the United States. The problem is worldwide. For example, in France eleven communities surrounding Orly Airport are seeking payment to soundproof 17 schools and five hospitals.[128]

Subsonic aircraft noise can be a challenge to the comfort and enjoyment of our homes; the future holds little promise of improvement. As a matter of fact, if the results of military supersonic overflights and sonic boom tests are indicative of things to come, the potential discomfort and loss in enjoyment of property will be much greater once commercial supersonic aircraft begin operating.

Sonic boom claims presented to the U.S. Air Force during the fiscal years 1956–1967 amounted to almost $20 million.[129] During this period $1.3 million was awarded in claims. Even though supersonic aircraft are in their infancy, claims for damage are appearing worldwide.

In southwestern France more than $400,000 has been awarded to satisfy the property damage claims of area residents because of flights of the Mirage IV Supersonic Bomber.[130] The British-French Concorde in just ten supersonic test flights over Britain's west coast has cost the government $3,120, and more damage claims are expected.[131]

If commercial SST's with their inherent booms become fully operational, the damage they might inflict would soar. According to CLASB, costs could exceed $3,000,000/wk. or nearly $160,000,000 yearly.[132] A more conservative estimate made by the Department of Interior is $35 to $80 million each year.[133] This figure includes the expected annual cost of repairing damaged homes and other properties, but not the amount required to process the claims and inspect the damages.

Over 65 million persons in an 80-mile-wide path could be exposed to at least eight daily booms.[134] These booms would in all probability be randomly distributed throughout the day, and no consistent sound pressure would occur. More than half the population sampled in all aircraft-community surveys felt they could never adapt to night booms regardless of their pattern. Their comfort and enjoyment of property would be severely curtailed.

As has been shown, it is not necessary to conjecture about the effect of future flights, since the existing supersonic military flights give a preview of what is to come. Not only man-made artifacts are affected—historical land formations and natural geographic features have also been reduced by this new technological device. Such damage handicaps man's comfort and enjoyment, especially in terms of recreational potential. It is entirely possible that supersonic flights over the snow regions of North America, where there are ski resorts, could precipitate snowslides and possibly avalanches.

Agricultural production may prove to be yet another environmental feature affected by noise. The growth rates of tobacco plants examined at an experimental station have registered a noteworthy decline under the influence of noise. As much as a 40-percent decrease in plant growth was found in a comparison with similar plants not exposed to a 100 decibel sound source for a period of two weeks.[135]

Whether damage involves broken windows, structural cracks, decreased property value, or a reduction in crop yield, annoyance of this type affects the comfort and enjoyment of environment. Noise-inflicted damage is as pervasive as is noise itself.

Interference of noise with messages vital to human well-being was described as being a Stage 1 and Stage 3 problem. Under the circumstances of Stage 3, hearing auditory messages is essential to efficiency (accomplishing a task, properly hearing instructions or orders, detecting emergency signals, etc.).

DISRUPTED SOCIAL INTERACTION

In Stage 4, the consequences of interference are not as severe as in Stage 3, because insuring the ability to be understood is less essential here to safety or individual well-being. Nonetheless, social communication and friendly interchange are essential to human living. Wherever such interchange takes place, it seems to be intruded upon by noise. Whatever the social activity—home entertaining, neighborhood fence talk, Kaffee Klatsches, street conversation, or television viewing—noise is ever present. It is increasing both in duration and in level.

Besides being a nuisance, noise may have more subtle effects. In situations where community noise is intense, it may affect interpersonal relations. "Misunderstandings" between persons, attributable to noise but not recognized as such, jeopardize friendships. For example, people become reluctant to visit or talk on the phone, especially long distance, to other people where high noise levels are a "way of life." This is especially so if one party is used to a quieter environment. When one or more methods of basic communicating (face-to-face, phone-to-phone) are masked by noise, the channels for social interaction become extremely limited. More than just a few households may be affected. Whole communities are vulnerable. Communities close to busy airports get little communication relief because of the noise caused by arriving and departing aircraft. The choice of times to communicate is not an individual decision, but rather one based on existing flight schedules.

HEARING LOSS

Physiological impairment to hearing occupies most of the attention of research. But there are secondary effects associated with hearing impairment that challenge human health. Such effects, behavioral in nature, rival physical disability. Hearing problems can bring about personality changes that alter patterns of behavior.

A person having a hearing loss may to others appear to be aggressive, out of place, or annoying due to his louder than normal conversational voice. He, on the other hand, while suffering this hearing loss may feel that others are not cordial or receptive to him. He may believe that his friends talk in a quiet, seemingly disinterested manner. In family situations, the consequences can be grave. According to one otologist, hearing loss is a common source of marital discord.[136]

The act of interpersonal communication involves culturally and socially prescribed distances and conversational levels to be employed when people talk face-to-face. If a person violates this sound-space (speaking-distance) relationship, not only is communication impeded, but interpersonal relations also change.[137]

IV

The Design of the Community Noise Survey

Our field survey was undertaken to observe at first hand how people respond to community noise, to what acoustical levels and sources of noise they are exposed, and ultimately to determine if community noise constituted a problem of public concern. By design the field investigation was "open-ended" and exploratory. It represented more than testing a series of hypotheses; it also offered an opportunity to learn about human responses to community noise previously unreported in the literature.

The survey was undertaken in metropolitan Philadelphia. It involved two political jurisdictions: the southwest portion of the city of Philadelphia and Tinicum Township, located in neighboring Delaware County (see Figure 4.1). This geographic area, containing 209,000 persons, was divided into twenty-five subareas. Monitoring stations were then established to administer a community opinion survey and obtain community noise readings. Five monitoring stations were established in Tinicum (see Figure 4.2); the remaining twenty stations were distributed throughout southwestern Philadelphia (see Figure 4.3).

Several criteria were used in the selection of these twenty-five locations. All the stations were located in residential areas, since the survey involved assessing the impact of community noise intruding upon the residential population. Consideration was also given to establishing stations near particular noise sources. For example, in Philadelphia, Station 2 was adjacent to the elevated-subway, Station 11 along a heavily traveled street. Similarly, Tinicum Station 4 was under the flight path for landing aircraft, Station 1 in the vicinity of a large

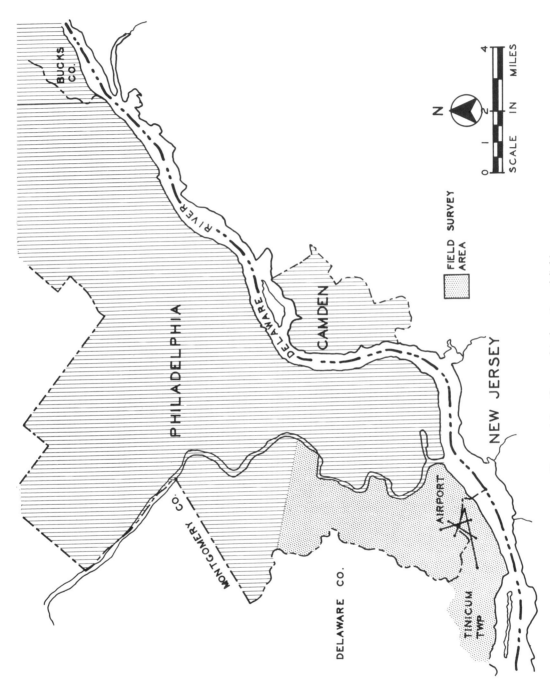

FIGURE 4.1. *Geographic location of field survey*

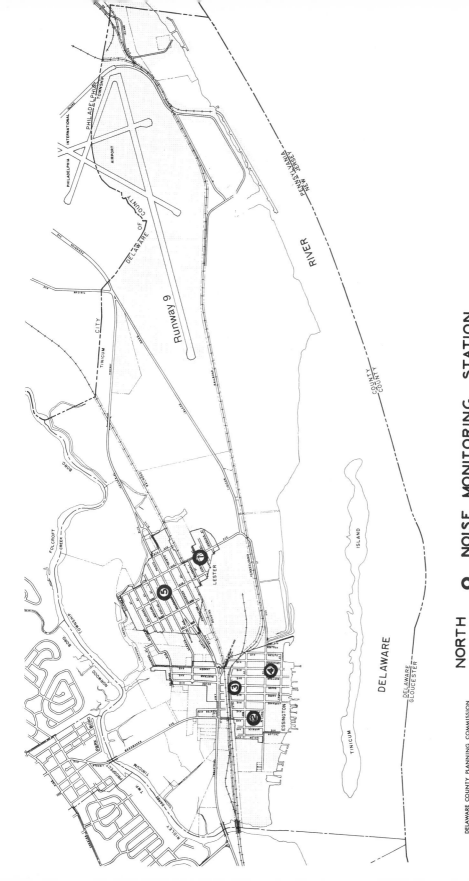

NORTH

○ NOISE MONITORING STATION

☐ DEVELOPED AREA

----- PROPERTY LINES

FIGURE 4.2. Noise monitoring stations: Tinicum Township, Pa.

DELAWARE COUNTY PLANNING COMMISSION
M E D I A   P E N N S Y L V A N I A

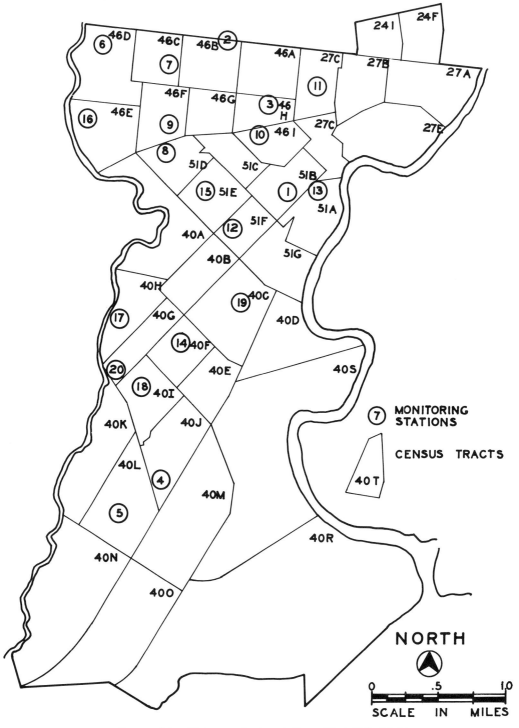

FIGURE 4.3. *Noise monitoring stations: West Philadelphia*

industrial facility. Still other stations were established where there was generally no discernible noise source. Beside noise sources, social and demographic factors were considered. The residential areas chosen represented a cross-section of urban society. High income-low income, white-Negro, low density-high density, little educated-highly educated, new residents-old residents, young adults-old adults, and home owners-apartment renters were among the population characteristics evaluated to achieve a representative distribution of monitoring stations. Much of the social information came from the 1960 U.S. Census of Population for Philadelphia as reported by census tract. With the exception of Tinicum, never more than one station was established in any census tract,[1] as indicated by Figure 4.3.

A total of 500 community opinion surveys was completed, and these constitute the sample size, or N. Twenty questionnaires were completed at each of the twenty-five stations.

Methodology: Social Survey
SAMPLE SIZE

Around each station an outline was drawn marking the area where interviews could be obtained. The general procedure was to administer the questionnaire to people living equidistant from the noise source or sources. This procedure insured that all residents interviewed at each station were exposed to community noise of similar intensity. For example, the sampling area drawn for Station 2 in Philadelphia was rectangularly shaped, and only those dwelling units facing the subway-elevated were queried. Usually each of the sampling areas contained approximately seventy-five residences.

SAMPLE METHOD

A quota method of sampling was used. Interviewers knocked on every door until twenty questionnaires were completed at each station. To avoid a bias in sampling and to complete the interview schedule as quickly as possible, the trained enumerators also worked on weekends. Generally their work hours were between the hours of 9 A.M. and 6 P.M. Because some stations had either a substantial refusal rate or several missed and vacant residences, the interviewers often revisited these areas. Just one questionnaire was permitted to be completed per dwelling unit, and it was filled out by only one adult (i.e., head of household or spouse).

Before the interviewers began their field work, they were given a two-day orientation, to acquaint them with the survey and its purposes. The first day in the field, after the orientation, was spent accompanying an interviewer already familiar with the questionnaire and thoroughly knowledgeable about census-taking procedures.

INTERVIEWING

A letter of identification was given to each interviewer indicating

TABLE 4.1.   Survey Sample Size

Survey area	Census population	Number of occupied households: Census	Number of occupied households: Sampled	Percent of occupied households: Sampled
Tinicum Twp.	4,375	1,300	100	7.63
West Philadelphia	204,502	65,088	400	0.62
TOTALS	208,877	66,388	500	0.76

that this survey was part of a research project jointly sponsored by the West Philadelphia Community Mental Health Consortium and the University of Pennsylvania's Institute for Environmental Studies. In introducing themselves, the interviewers told the residents they were administering a community opinion survey to determine how people feel about the environment in which they live. To avoid any possibility of bias, the *survey was not introduced as a noise survey*.

When invited into a dwelling, the interviewer sat down with the respondent and recorded answers to the survey questions. With only a few exceptions, the questionnaire was filled out in the interviewer's presence. The time required to complete a questionnaire averaged fifty minutes.

SAMPLE
COMPOSITION

Five hundred households received and completed the questionnaire in a survey area containing 66,388 occupied dwelling units. The percentage of those dwelling units sampled represented 0.76 percent of the total area population (see Table 4.1). Comparing West Philadelphia and Tinicum Township, the latter group had the larger percentage of households sampled, while West Philadelphia had the larger absolute number.

COMPARATIVE
SOCIAL
CHARACTERISTICS:
SURVEY SAMPLE
AND CENSUS

The greatest discrepancy in characteristics between the actual census population and the population sampled by this survey concerned the number of males queried, and the number of females in the labor force. The differences were not major, however. Table 4.2 shows that, proportionately speaking, 30.39 percent of those interviewed were males, although males constitute 47.41 percent of the area's population. Although 42.84 percent of all females (14 years and over) are in the labor force, those interviewed represented only 23.30 percent (see Table 4.8). Other than these two exceptions, it appears that the sample was representative (see Tables 4.2–4.9).

Sex of Sample

TABLE 4.2.

Category	Sample		Census	
	Number	Percent	Number	Percent
Male	148	30.39	99,035	47.41
Female	339	69.61	109,821	52.59
TOTAL	487	100.00	208,856	100.00

Race of Sample

TABLE 4.3.

Category	Sample		Census	
	Number	Percent	Number	Percent
White	341	68.20	152,746	73.13
Negro	153	30.60	55,170	26.41
Other	6	1.20	960	.46
TOTAL	500	100.00	208,876	100.00

Age Distribution of Sample

TABLE 4.4.

Category	Sample		Census	
	Number	Percent	Number	Percent
20–29 yrs.	90	18.75	30,886	21.66
30–39 yrs.	99	20.63	27,371	19.20
40–49 yrs.	77	16.04	25,818	18.11
50–59 yrs.	101	21.04	23,974	16.81
60–69 yrs.	71	14.79	19,478	13.66
70+ yrs.	42	8.75	15,068	10.56
TOTAL	480	100.00	142,595	100.00

Housing Tenure

TABLE 4.5.

Category	Sample		Census	
	Number	Percent	Number	Percent
Owner occupied	365	74.64	37,265	57.67
Renter occupied	124	25.36	27,345	42.33
TOTAL	489	100.00	64,610	100.00

Average Household Size

TABLE 4.6.

Category	Sample	Census
Persons per Dwelling unit	3.51	3.14

TABLE 4.7.   Income of Residents

Category	Sample		Census	
	Number	Percent	Number	Percent
Less than $5,000	94	24.04	19,200	37.99
$5,000–$9,999	196	50.13	24,956	49.38
$10,000–$14,999	67	17.14	5,246	10.38
$15,000+	34	8.69	1,138	2.25
TOTAL	391	100.00	50,540	100.00

TABLE 4.8.   Employment by Sex

Category	Sample		Census	
	Number	Percent	Number	Percent
Females in labor force	79	23.30	36,625	42.84
Females not in labor force	260	76.70	48,874	57.16
TOTAL	339	100.00	85,499	100.00
Males in labor force	80	54.06	55,700	43.24
Males not in labor force	68	45.94	73,128	56.76
TOTAL	148	100.00	128,828	100.00

TABLE 4.9.   Adult Educational Attainment

Category	Sample		Census	
	Number	Percent	Number	Percent
No school completed	1	.34	2,362	3.59
Elementary completed to 8th	70	23.81	25,762	39.22
High school completed	178	60.54	29,353	44.69
College completed	45	15.31	8,213	12.50
TOTAL	294	100.00	65,690	100.00

The questionnaire (see Appendix A) prepared for the field survey contained three sections. Section I dealt with the general residential environment. No questions were asked in Section I concerning noise, and the emphasis was on how well the respondents liked the physical-social environment where they lived. Several questions sought to measure the degree of identification with and attachment to their local environment. The extent of community friendships, community participation, and patterns of physical mobility were some of the other areas investigated. Human response to community noise was the principal subject of Section II. The *residential environment* was the reference point for opinions about noise heard, rather than the *city at large,* which has been the case in many other noise surveys. In this way the task of comparing attitudes to existing noise levels was fixed in time and space. Opinions could then be correlated with noise levels and sources.

The usual questions about noise nuisance were asked: "What are the types of noises that annoy you?" "Does noise annoy you during the following list of activities?" "Is there any time of the year when noise disturbs you most?" but there were also other questions, concerning the effect of noise upon human health ("Do you feel any different after being exposed to noise?" "Do you consider noise harmful to your health and well-being?"), its economic impact, and the citizens' role in abating noise. The final section stressed demographic questions. Many of these questions were purposely identical with those used by the U.S. Bureau of Census, in order to determine the validity of the sample. Equally important was the chance to compare, by means of statistical analysis, several social variables (age, sex, race, education, employment, income, etc.) with human responses to noise.

The twenty-five monitoring stations served a dual purpose, delineating the locations for conducting the social survey and obtaining readings of existing community noise levels. To measure noise during representative periods of city activity, a 24-hour, seven-day sampling schedule was devised. Two criteria were used for establishing the sampling periods: selecting four general time periods of the day (morning, afternoon, evening, and night) and choosing periods of high, medium, and low mobility. These criteria appear to be key factors in community noise variation, upon analysis of several studies reported in the literature.

Population mobility was derived by averaging traffic volume by hour for all monitoring stations during a 24-hour-a-day cycle (see Figure 4.4). Most transportation modes that could be potential community noise sources were included in the traffic volume estimate: aircraft, bus, trolley, subway, trains, and vehicles (cars-trucks). Gener-

QUESTIONNAIRE

Methodology: Acoustic Survey

SAMPLE METHOD
General Community Noise Survey

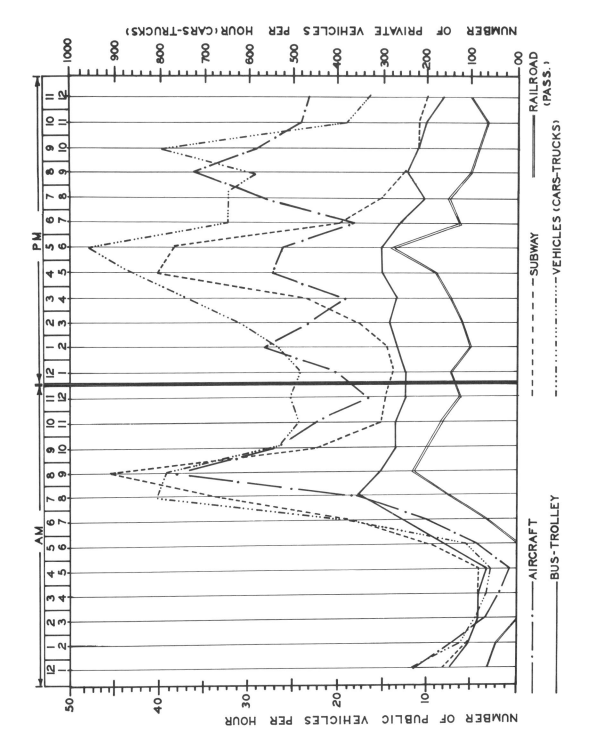

FIGURE 4.4.   *Average weekday traffic volume per hour in field study area: West Philadelphia–Tinicum Township*

ally speaking, low mobility was found between 11 P.M. and 6 A.M. and medium mobility from 6 A.M. to 7 A.M., 10 A.M. to 3 P.M., and 7 P.M. to 11 P.M.; while the highest mobility occurred from 7 A.M. to 10 A.M. and from 3 P.M. to 7 P.M.

When each hour of the day was classified according to low, medium, or high mobility, six sampling times were assigned for taking noise measurements: low mobility 11 P.M.–2 A.M. and 2 A.M.–6 A.M.; medium mobility 10 A.M.–3 P.M. and 7 P.M.–11 P.M.; high mobility 7 A.M.–10 A.M. and 3 P.M.–7 P.M. (see Figure 4.5). Noise samples were taken at each monitoring station on four different days (three weekdays and one weekend day). During each of the three weekdays, measurements were taken during the six different time spans. On the weekends, each monitoring station was visited five different times and noise measurements were recorded. Ultimately all the designated monitoring stations were sampled, with each station visited four different days of the week. A total of 23 visits were made at each station, 18 during the weekdays and 5 on the weekends. During each visit to a monitoring station, a noise sample was recorded for a period of five consecutive minutes.

This was one of two sample methods utilized in the field survey. This first method was used to determine the overall community noise level. Individual noise sources were not isolated in any way.

The second method involved identifying and then measuring the principal individual components of community noise found at these monitoring stations.

First a listing of specific sources of community noise by station was compiled (Figure 4.6). Then a preliminary field check was undertaken, using a sound level meter to determine what noise sources should be surveyed. Results of the social survey provided a listing of noises considered annoying by the inhabitants. The final list included the following identifiable noise sources: aircraft, bus, subway, trolley, railroad, vehicles (cars, trucks), industry, and children.

Survey of Specific Community Noises

For each source a sample was taken. The distribution of each of these noise sources and their average sound pressure levels measured in dBA are presented in the next chapter (see Figures 5.8–5.13).

General Radio Company provided most of the acoustical equipment required to obtain the noise readings. The field survey instruments (see Figure 4.7) consisted of a Sound Level Meter (Type 1551-C), Octave-Band Noise Analyzer (Type 1558), Data Recorder (Type 1525-A), Graphic Level Recorder (Type 1521-B), inverter, microphone stand, windscreen, and calibrator. It was later necessary to augment this equipment with a Brüel and Kjaer Sound Level Analyzer (Type 2203) and Level Recorder (Type 2305). All the measuring instruments meet

FIELD SURVEY INSTRUMENTS

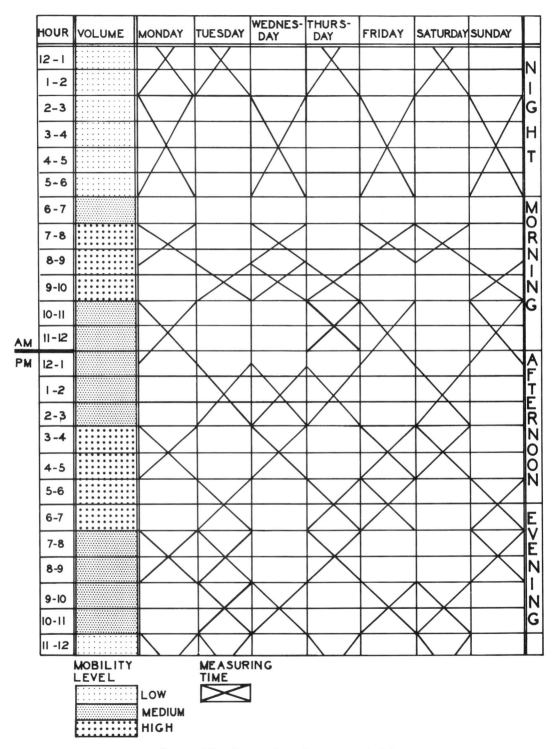

FIGURE 4.5.   *Community noise survey schedule*

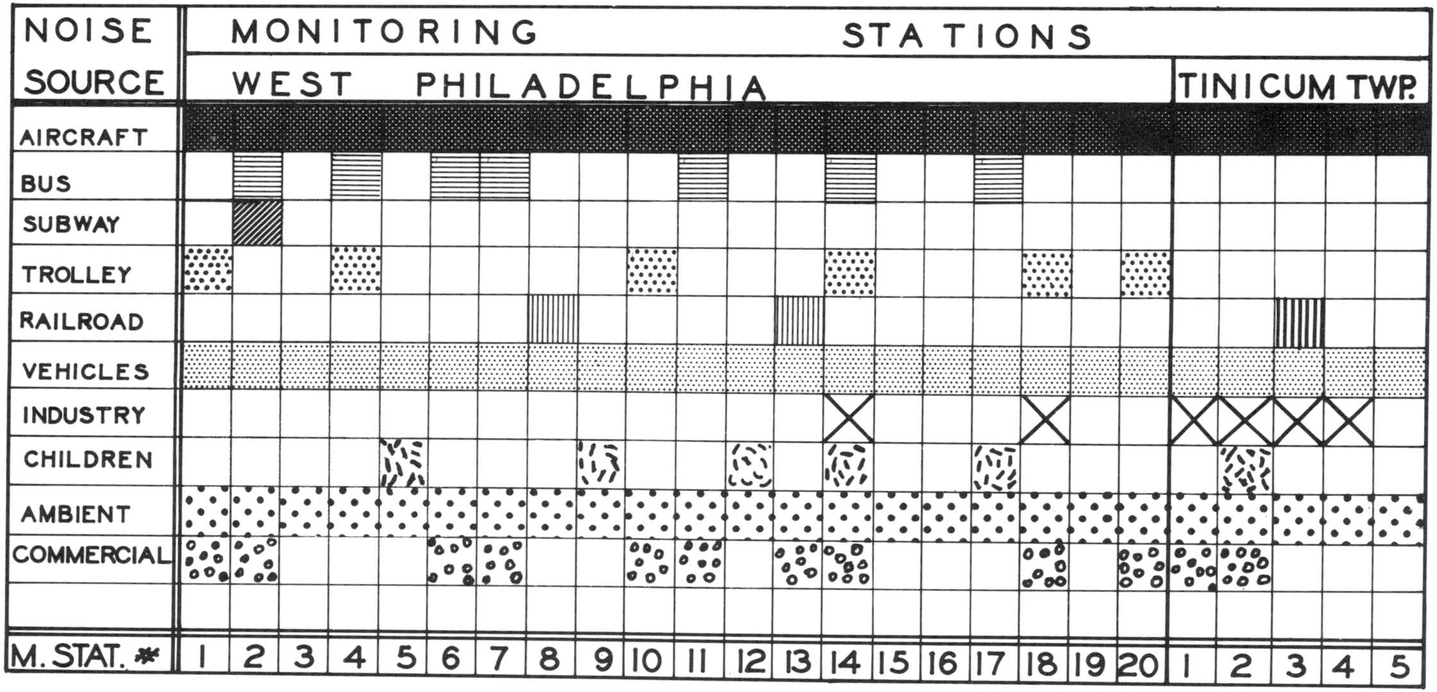

NOISE SOURCE | MONITORING STATIONS

WEST PHILADELPHIA | TINICUM TWP.

AIRCRAFT
BUS
SUBWAY
TROLLEY
RAILROAD
VEHICLES
INDUSTRY
CHILDREN
AMBIENT
COMMERCIAL

M. STAT. # | 1 2 3 4 5 6 7 8 9 10 11 12 13 14 15 16 17 18 19 20 | 1 2 3 4 5

FIGURE 4.6. *Community noise sources by monitoring station: field study area (West Philadelphia–Tinicum Township)*

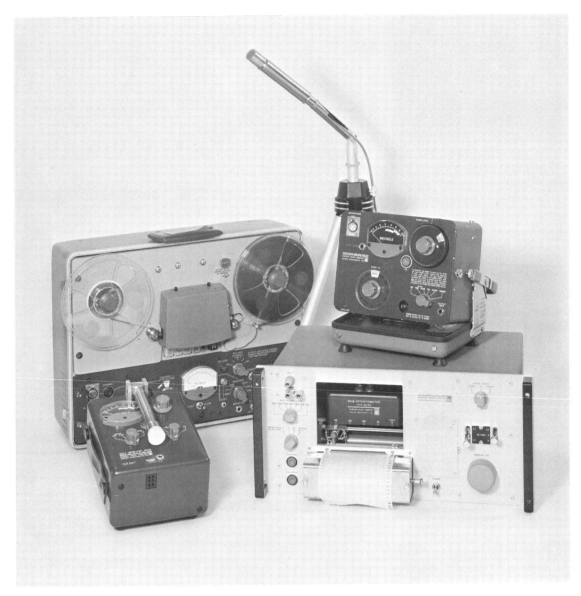

FIGURE 4.7. *Community noise field survey equipment*

the American Standards Institute (ANSI) standards for electronic accuracy.

MEASUREMENT PROCEDURES

Several guidelines were used to secure consistent and accurate observations. Adverse weather conditions were avoided so as not to bias the data. During periods of rain, sleet, snow, hail, and/or high winds, no measurements were taken. Field survey work occurred only under "normal" weather conditions, including times when the pavement was free of any precipitation. To minimize possible distortion caused by winds in excess of five miles per hour, a windscreen was fitted over the microphones for all field measurements.

Before and following the series of field measurements at each monitoring station, an acoustical calibrator checked amplifier gain and sound level response. The microphone attached to the recording device was placed so as to avoid reflecting surfaces, such as walls, landscape barriers, and so on. A minimum distance of ten feet from such distorting features was always maintained. No obstructions were permitted between the noise source and the instruments measuring that noise. Measurement procedures were identical in the general noise survey and the survey of specific noises, with one exception. The distance from the noise source varied.

In this survey measurements for every monitoring station were taken at the curbside of the roadway. The survey instruments were contained in a station wagon with the microphone of the sound level meter mounted on the roof. This microphone was then attached by cable to a graphic level recorder, powered by an inverter, both located inside the vehicle. Generally speaking, the distance between the microphone location, where measurements were taken, and private property averaged 10–15 feet. Although it might have been preferable to obtain noise level readings directly at the property line, the difference in readings was considered minor. By locating at the curb there was the advantage of concealing the instruments and therefore minimizing attention in the area.

General Community Noise Survey

For conducting this survey a horizontal distance of 15 to 100 feet from the noise source (with the exception of aircraft) was used, since this distance is a commonly accepted standard. In order to obtain the proper distance for these measurements, a microphone stand was set up, and readings were observed in unobstructed open areas.

Survey of Specific Noises

Based upon the sampling procedure already detailed, the analysis involved computing median and inter-quartile ranges of general community noise by time of day (i.e., the noise level occurring during the weekday and weekend, and for the total week). Morning, afternoon, evening, and nighttime community noise were determined as well. The intensity of noise as a function of mobility was given equal consideration. Geographic breakdowns were made for each monitoring station and for the two political areas sampled, the City of Philadelphia and Tinicum Township.

ANALYSIS OF ACOUSTICAL SURVEY
General Community Noise Survey

For every category of community noise (subway, aircraft, vehicle, etc.) an average was computed by using the peak sound pressure level

Survey of Specific Community Noise

of the noise as the point of measurement. Since time and mobility did not alter the noise level of each source, only one average was used.

Community Description

LAND USE

This portion of the metropolitan Philadelphia area, although primarily residential, contains a large number of industrial facilities. Along the western side of the Schuylkill River, a nearly continuous strip of industrial activity reaches southward to the Philadelphia International Airport (see Figure 4.8). Adjoining the City of Philadelphia on the southwest is Tinicum Township, which houses over 5,000 persons. Tinicum lies between the airport on the east and heavy industrial facilities on the west extending unbroken along the Delaware River to Wilmington, Delaware (see Figure 4.9). These industries, which include ship-building installations, refineries, foundries, and other manufactures, draw most of the Tinicum labor force and a sizable number of workers from the southern half of West Philadelphia.

Overall, the residential area is less densely populated in Tinicum than in Philadelphia. The Township also contains a much higher proportion of detached single-family dwelling units. Although Tinicum is a very old settlement, large areas of uninhabited land surround its two adjoining communities, Lester and Essington (see Figures 4.10 and 4.11). In the southern part of West Philadelphia there is a sizable undeveloped area. These lands have a marginal residential potential because of poor geological conditions and neighboring industry.

ACTIVITY PATTERNS

The survey area is a major traffic generator, with the airport, two military installations, and many industries in its vicinity. Another employment center is University City situated in the northeastern part of West Philadelphia. This area contains several educational institutions and research-oriented businesses.

Despite the fact that the Township and Philadelphia share a common boundry, the activity patterns of the two populations are very different. The Tinicum work force does not as a rule rely on the city for employment. The majority either have jobs within the Township or commute to the north or west rather than the east. Consequently, Philadelphia is alien to a great many. Social and recreational patterns are also localized. Even though many once lived in Philadelphia, they have isolated themselves since relocating. The feeling is mutual, however. Most of the people interviewed in West Philadelphia knew little about the Township. If known at all, it was vaguely referred to as being "that place somewhere near the airport."

PHYSICAL ENVIRONMENT

The physical environment where our survey was conducted is typical of most industrially developed, highly populated urban areas. Air pollution is a common problem. The air has a high level of pollutancy containing both gaseous and particulate matter. Open bodies of

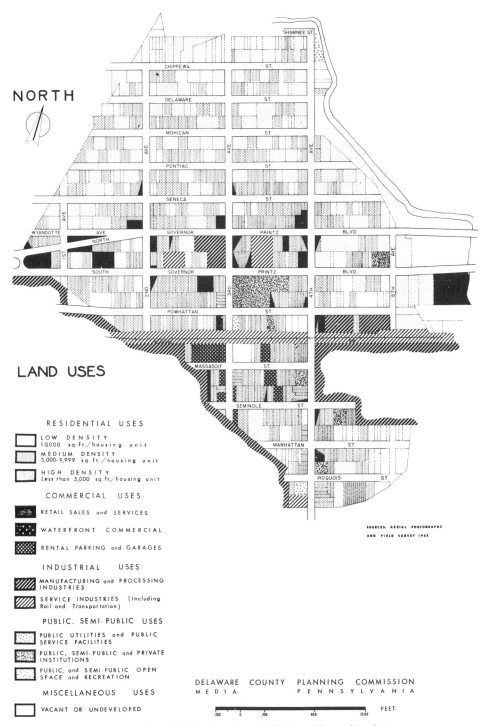

NORTH

LAND USES

RESIDENTIAL USES

LOW DENSITY
10,000 sq.ft./housing unit

MEDIUM DENSITY
5,000-9,999 sq.ft./housing unit

HIGH DENSITY
Less than 5,000 sq.ft./housing unit

COMMERCIAL USES

RETAIL SALES and SERVICES

WATERFRONT COMMERCIAL

RENTAL PARKING and GARAGES

INDUSTRIAL USES

MANUFACTURING and PROCESSING INDUSTRIES

SERVICE INDUSTRIES (Including Rail and Transportation)

PUBLIC, SEMI-PUBLIC USES

PUBLIC UTILITIES and PUBLIC SERVICE FACILITIES

PUBLIC, SEMI-PUBLIC and PRIVATE INSTITUTIONS

PUBLIC, and SEMI-PUBLIC OPEN SPACE and RECREATION

MISCELLANEOUS USES

VACANT OR UNDEVELOPED

DELAWARE COUNTY PLANNING COMMISSION
MEDIA   PENNSYLVANIA

FEET.

FIGURE 4.11. *Existing land use in Tinicum Township: Lester*

water (the Schuylkill and Delaware Rivers), because of intense industrial usage (refineries, manufacturing plants, etc.) have only limited recreational use. Several boating and yacht clubs, although still based in Tinicum along the Delaware River, because of the prevailing conditions no longer enjoy the popularity they once had. Openness is one of the few amenities still offered, at least in Tinicum. For most of the residents, particularly those in West Philadelphia, travel is becoming the one way to find a better environment.

GENERAL DEMOGRAPHY

Social Composition

In 1960, the survey area was predominantly white, with one in four inhabitants being non-white. For Tinicum alone, the 1960 census figures indicate the absence of a non-white population. Best estimates today suggest that in West Philadelphia approximately 40 percent of the population are non-whites (nearly all of them Negro), while Tinicum is about one percent non-white. The white group is considerably older than the non-white group. This entire survey area has a sizable aging and retired population of long-term residents. By contrast, most of the Negroes began to settle in West Philadelphia during the early fifties. They were younger and more highly educated Negroes than the national average.

It is difficult to type West Philadelphia today because the area is heterogeneous, but Tinicum is less so. The Township is mainly a lower-middle, white working-class community.

Housing

Over half the survey population live in owner-occupied housing. Since there is a considerable number of community-unrelated individuals in West Philadelphia—attending college, etc.—the number of renter-occupied units approaches 40 percent. In Tinicum approximately four out of every five families occupy their own residences. Housing conditions remain average to good, with only minor dilapidation and a small number of deteriorating structures. There are a few pockets of substandard housing in West Philadelphia. Household size averages slightly above three persons per dwelling unit.

Other Data

On the whole the amount of education completed was equivalent to a high school diploma, though approximately one-fifth of those who completed high school also completed college. The level of education is reflected somewhat by type of employment. Jobs are primarily in the blue collar classification (including craftsmen, operatives, domestics, service workers, and laborers). A final demographic factor is income, which is related to both education and employment. According to 1960 census information, median family income in West Philadelphia was approximately $5,800, while in Tinicum the median was nearly $1,000 higher, or $6,800. Contributing to lower the West Philadelphia average was the lower non-white median family income of $5,300.

V

The Metropolitan Philadelphia Noise Survey: Analysis of Sound and Its Sources

Seven different noise sources were surveyed, all of them means of transportation. Three were public conveyances operated by the Southeastern Pennsylvania Transportation Authority: subway-elevated, trolley, and bus. The other four were privately operated conveyances: automobile, truck, train, and airplane.

A set of similar distances was used to obtain overall sound pressure level measurements (dBA) for each noise source with the exception of aircraft. Measurements were taken at distances of 15, 25, 50, and 75 feet. The size of the samples varied. All sites chosen for these measurements represented the most open land areas available—so chosen to minimize sound reflection. In every instance the vehicles were operating close to normal city speeds (30–35 mph), a fact that made valid comparisons among all sources possible. In some cases, the noise levels observed were not peak noise levels. For example, the peak noise level of a diesel truck occurs during maximum low-gear acceleration rather than at normal city cruising speeds.[1]

Different procedures were used in measuring aircraft noise. First of all, the survey examined scheduled air carrier operations at two general

Survey of Specific Noise Sources

locations. One location was established adjoining the Philadelphia International Airport, to monitor take-off and landing noises. This location consisted of nine monitoring points, all within Tinicum Township, where aircraft altitude was consistently below 1,500 feet. In West Philadelphia, the second location, where aircraft were at altitudes above 2,000 feet, twenty monitoring stations were established.

General Noise Levels
SURFACE
TRANSPORTATION

The subway-elevated was consistently the noisiest of all six surface transportation sources (see Figure 5.1). At a distance of fifteen feet the sound level averaged 97 dBA. Passenger-commuter train noise ranked second, followed closely by truck noise ranked third, and trolley noise ranked fourth. The difference between these last two was slight—less than two decibels. (At fifteen feet the level for truck and trolley was 88 and 87 dBA respectively.) Had the truck sample included only tractor trailers, the difference between the two would have been greater. However, all sizes of trucks, from light to heavy, were sampled. Bus noise averaged approximately six decibels less than truck and trolley noise. Automobiles generated the lowest noise level: 78 dBA at fifteen feet and 63 dBA at seventy-five feet.

AIR
TRANSPORTATION

The intensity of commercial aircraft noise was markedly higher in Tinicum Township than in West Philadelphia. Despite a smaller sample taken within the city, it appears that for jet aircraft the average difference amounted to 16 dBA (95 dBA as opposed to 79 dBA—see Table 5.1).

Type and size of aircraft affect the noise output. Piston powered planes (including those classified as turboprop) operate more quietly than do planes powered by jet engines. In our survey, jet planes monitored in Tinicum averaged 15 dBA higher than planes with reciprocating engines flying over the same area. The difference was less in West Philadelphia because jet aircraft were at a higher altitude than prop aircraft. Aircraft size, too, is important: aircraft with the larger number of engines generate greater noise intensity. This difference is not very

TABLE 5.1.   Sound Level of Aircraft by General Type and Location

	Tinicum Township		West Philadelphia	
Aircraft type	*Sound level*	*Sample size*	*Sound level*	*Sample size*
Jet engine	95 dBA	333	79 dBA	111
Reciprocating engine	80 dBA	96	73 dBA	14
TOTAL		429		125

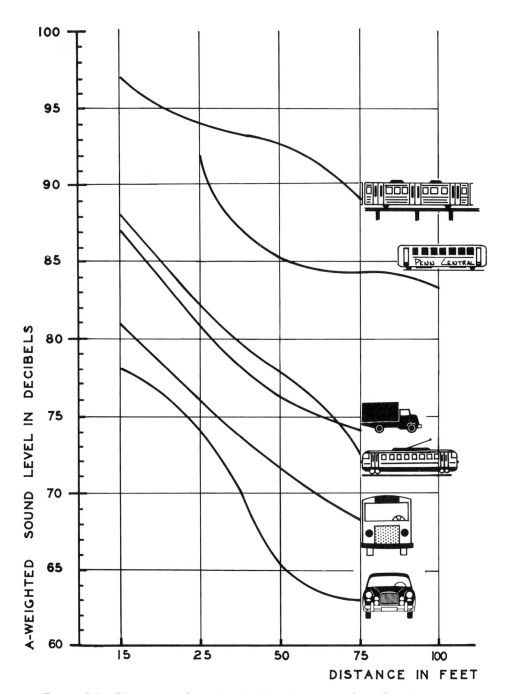

FIGURE 5.1. *Distance as a factor in noise intensity: survey of specific noise sources*

noticeable among two- and four-engine piston planes; it averages approximately 2 dBA (see Figure 5.2). More significant are the comparisons of different sizes of jet aircraft landing at the airport. Three-engine jets were 3 dBA greater than two-engine ones, and 5 dBA lower than four-engine jets. The difference in the noise level between two- and four-engine jets was thus 8 dBA. During takeoff, when airplanes are under full power, the difference in the number of engines is even more noticeable (see Figure 5.3). The largest jets were found to have a noise

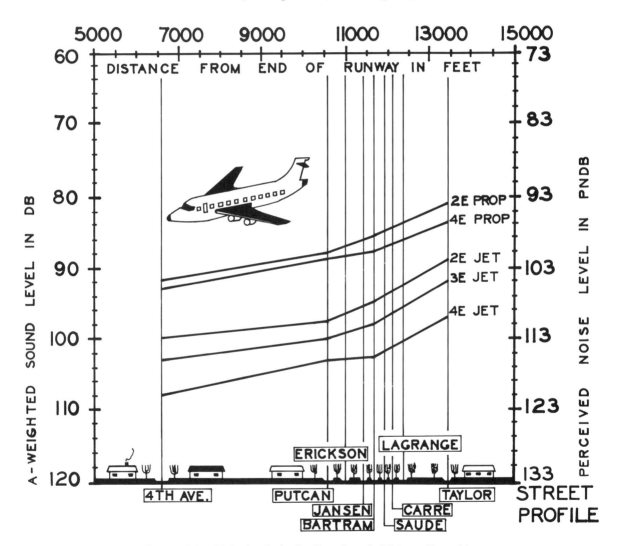

FIGURE 5.2. *Noise levels for landing aircraft: Tinicum Township*

level approximately 8 dBA higher than three-engine jets and 12 dBA higher than the smaller two-engine jets.

These findings are similar to those obtained in other research work. The firm of Bolt, Beranek, and Newman has developed a series of contour maps comparing the noise levels of various aircraft types during takeoff and landing (see Table 5.2 and Figures 5.4–5.6).[2] The noise levels are expressed in PNdB, but as mentioned in Chapter II, can be converted to dBA by subtracting 13. Figure 5.2 indicates that the variation

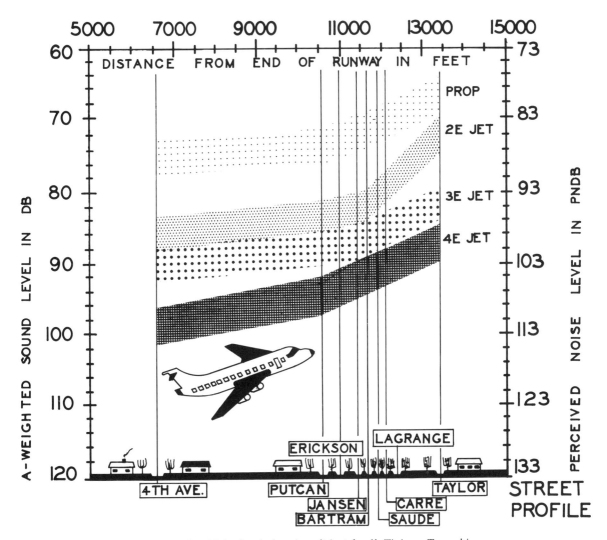

FIGURE 5.3. *Noise levels for aircraft in takeoff: Tinicum Township*

TABLE 5.2.
Chart for Selecting Noise Contours for Civil Aircraft

| | OPERATION | | | | | |
| AIRCRAFT TYPE | TAKEOFFS | | LANDINGS | | RUNUPS | |
	Contour set	Corrections to contour, PNdB	Contour Set*	Corrections to contour, PNdB	Contour set	Corrections to contour, PNdB
Large turbojet transports — Boeing 707 and 720, Convair 880, Comet; Douglas DC-8 — Trips under 2000 mi.	1A	0				
Large turbojet transports — Boeing 707 and 720, Convair Comet; Douglas DC-8 — Trips over 2000 mi.	1B	0	3B	0	6	0
Large turbofan transports — Boeing 707 and 720, Convair 990, Doublas DC-8 — Trips under 2000 mi.	1A	−5				
Large turbofan transports — Boeing 707 and 720, Convair 990, Douglas DC-8 — Trips over 2000 mi.	1B	−5	3B	0	12	0
Medium and short range turbofan transports — Boeing 727, Douglas DC-9	9A	0	10A	0	12	0
BAC 1-11	9A	+5	10A	−5	-	-
Medium range turbojet transports — Caravelle (3 and 6 series)	9A	+5	10A	0	6	0

Category	Aircraft						
Business turbojets	Jet Commander 1121, Lear Jet 23, Hawker Siddeley D.H. 125	9A	0	10A	−5	8	0
	Lockheed Jet Star, North American Sabreliner	9A	−5	10A	−10	8	−5
Business turbofans	Dassault Falcon	9A	−5	10A	−10	12	
Four engine piston transports	Douglas DC-4, -6, -7, Lockheed Constellation	4	0				
Four engine turboprop transports	Lockheed Electra	4	−5	3A	0		
Two engine piston transports	Douglas DC-3, Convair 240, 340, 44, Martin 202, 404	9B	0	10B	0		
Two engine turboprops	Fairchild F-27, Grumman Gulfstream	9B	−5	10B	0		
Two engine light piston business aircraft**	Aero Commander, Beech 18 series, Cessna 310 series, Piper Apache and Aztec, etc.	9B	10	10B	−10		

*Contour Set 11 may be used instead of Contour Set 10 for estimating noise exposure at airports not possessing instrument landing facilities or where only a very small proportion of instrument approaches are made.

**Two engine piston aircraft from 3,500 to 10,000 lbs. gross weight.

Source: Bolt, Beranek, and Newman, Inc., "Procedures for Developing Noise Exposure Forecasts Areas for Aircraft Flight Operations," prepared for the Federal Aviation Agency, August, 1967.

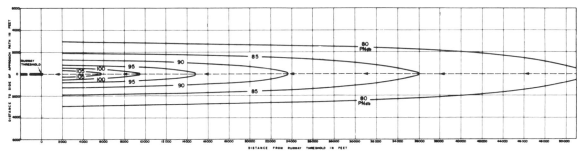

(A) FOUR-ENGINE PISTON AND TURBOPROP AIRCRAFT.

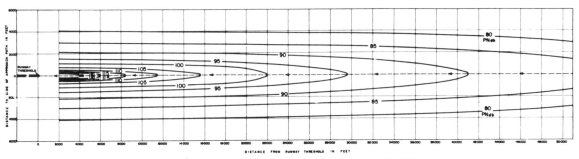

(B) TURBOJET AND TURBOFAN AIRCRAFT.

FIGURE 5.4. *Perceived noise level contours for civil and military landings*

between non-jet aircraft (four-engine piston and turboprop) and jet aircraft (turbojet and turbofan) in landing is 10 to 15 dBA, depending upon the distance from the runway threshold (point where aircraft first makes contact with runway). For this same type of aircraft measured in Tinicum the difference ranged from 8 to 16 dBA.

Noise Levels and Distance

Noise diminishes with distance. In a far field, as the distance is doubled the sound level is reduced by six decibels[3] (see Figure 5.7). This phenomenon is explained by the inverse square law of physics and refers to an "open field" condition (a field free of objects). Under such conditions, a motorcycle producing 90 dBA at twenty-five feet produces 84 dBA at fifty feet and 78 dBA at one hundred feet. The appropriateness of the inverse square law was tested in an urban environment where an open field condition does not often exist. The effect of both horizontal and vertical distance from a noise source was examined, since the survey of specific noises covered both surface and air transportation sources.

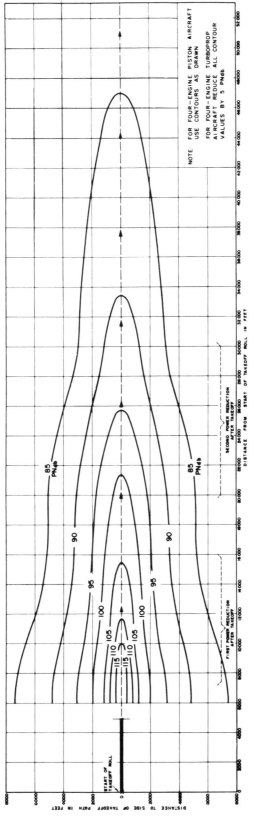

FIGURE 5.5. *Perceived noise level contours for takeoffs for four-engine piston and turboprop aircraft*

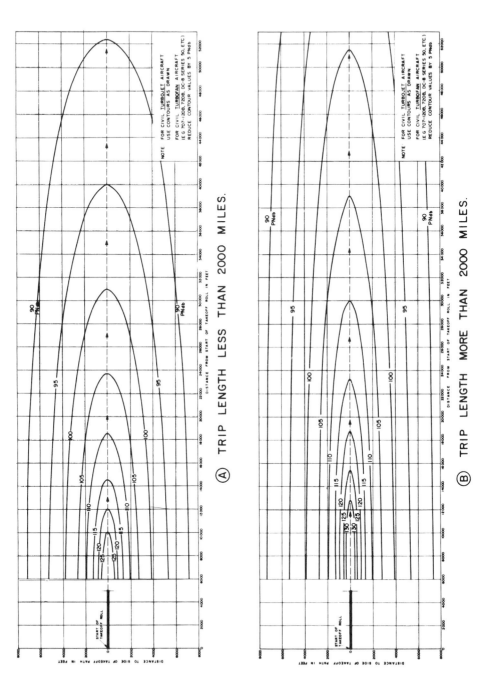

FIGURE 5.6. *Perceived noise level contours for takeoffs of civil jet transports*

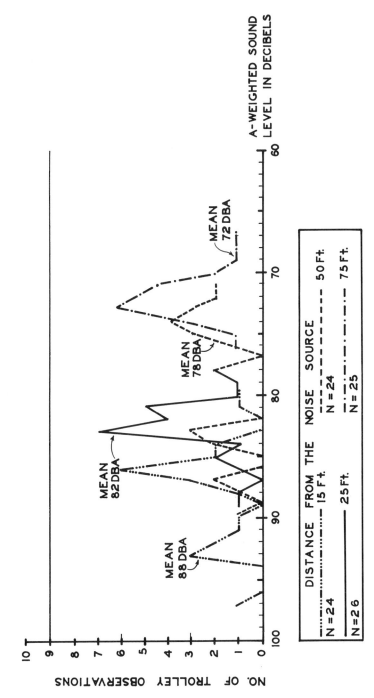

FIGURE 5.10. *Survey of specific noise sources: trolley*

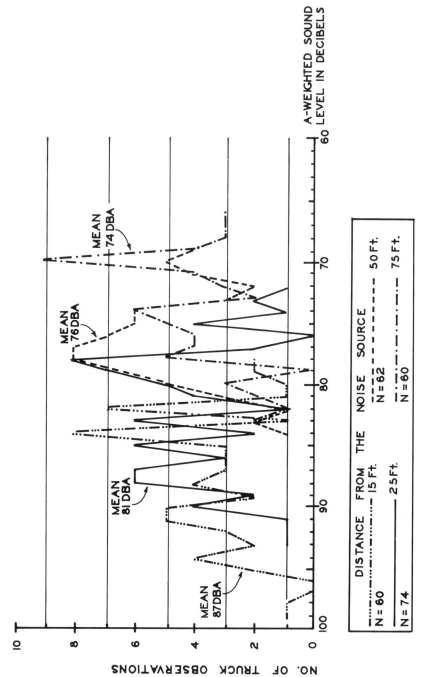

FIGURE 5.11. *Survey of specific noise sources: truck*

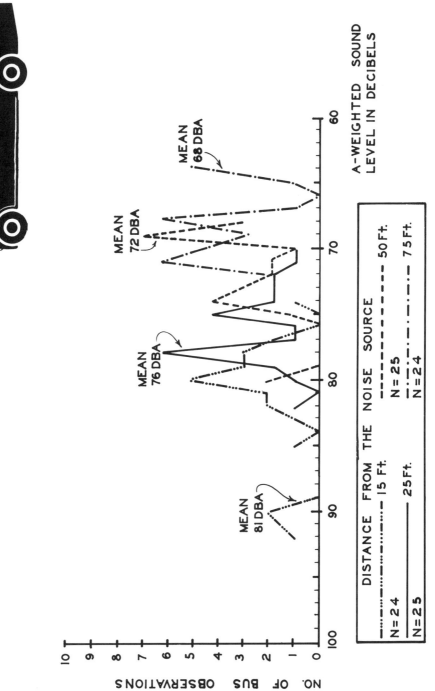

FIGURE 5.12. *Survey of specific noise sources: bus*

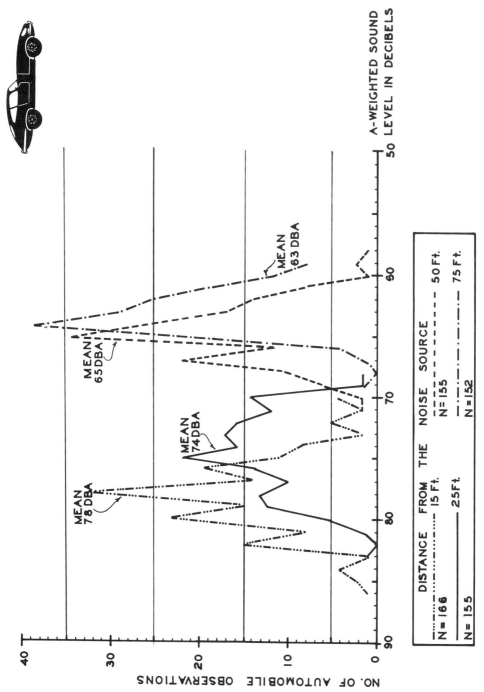

FIGURE 5.13. *Survey of specific noise sources: automobile*

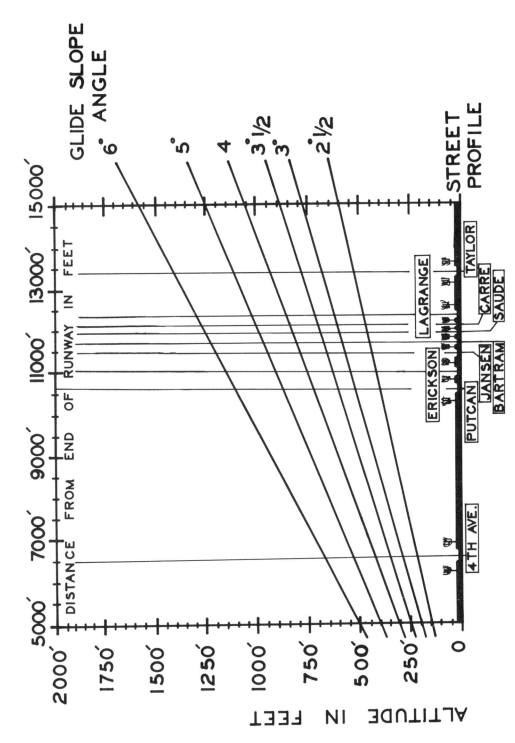

FIGURE 5.14. *Landing profiles for several glide slope angles: Tinicum Township*

TABLE 5.4.   Noise Level as a Function of Vertical Distance: Air Transportation

Noise source: Aircraft	Vertical distance from noise source			
	275 ft. altitude	4th Avenue	500 ft. altitude	Lagrange Avenue
4 engine jet	108 dBA		101 dBA	
3 engine jet	103 dBA		97 dBA	
2 engine jet	100 dBA		94 dBA	
4 engine prop	94 dBA		87 dBA	
2 engine prop	92 dBA		85 dBA	

closed a group of buildings are, the lower the noise level (see Figure 5.15). They suggest that buildings planned in a parallel pattern can be 3 decibels quieter than a semi-open arrangement of buildings, and 6 decibels quieter than an enclosed series of buildings. Building density and height are another consideration. Again, no substantial studies are available. To determine the effect of building height on noise propagation at street level, three different sites were compared in the study area

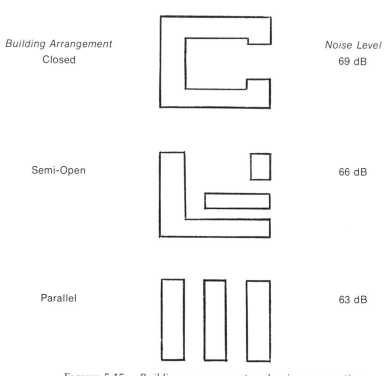

Building Arrangement   Noise Level
Closed   69 dB

Semi-Open   66 dB

Parallel   63 dB

FIGURE 5.15.   Building arrangement and noise propagation

Building Height and Noise Propagation

TABLE 5.5.

Noise source	High (3 story)	Sample size	Medium (2 story)	Sample size	Low (1 story)	Sample size
Elevated subway	98 dBA	10	96 dBA	10	94 dBA	15
Bus	86 dBA	10	85 dBA	10	76 dBA	24
Trolley	85 dBA	20	84 dBA	20	82 dBA	24
Truck	84 dBA	50	82 dBA	35	81 dBA	60
Auto	77 dBA	51	77 dBA	51	74 dBA	155

(see Figure 5.16). The low density street chosen had single-story build-ings while the medium had two-story and the high, three-story build-ings. Five types of ground transportation were measured at a distance of 25 feet at each of the three sites. It was found that building height does affect the intensity of noise generated at street level. Usually the variation in the noise level was greater between one- and two-story buildings than between two- and three-story buildings (see Table 5.5). The net difference between one- and three-story buildings varied from 3 dBA for trucks to 10 dBA for buses. It can be concluded that surface modes of transportation operated under similar conditions can generate a noise level of 3 to 4 decibels higher in a street of three-story buildings than in a street of single-story buildings. Had it been possible to com-pare streets with much higher buildings, a more complete picture of the relation of building height to noise density would be available.

The mean noise level for all our monitoring stations during the sampling time from 7:00 A.M. to 11:00 P.M. was 57 dBA (see Figure 5.17). The average minimum noise level recorded at these stations was 47 dBA, while the average peak was 81 dBA (see Figures 5.18–5.19). Considerable variation was observed among the various stations, particularly in West Philadelphia. In the city the highest mean noise level, 65 dBA, was recorded at Stations 1 and 20 while Station 3 recorded the lowest; it was 17 decibels lower, or 48 dBA. There appears to be a significant variation in residential area noise levels, even when the monitoring stations are close to one another. Consequently, *generaliza-tions should be avoided in classifying the residential noise level of a city or portions thereof.* Interestingly, some sections of Philadelphia were nearly as quiet as rural areas, while other sections were as noisy as many industrial plant operations. In Tinicum, however, the differences among stations were much smaller; there the largest variation was only 5 dBA, compared to 17 dBA in Philadelphia.

Although the particular mode of transportation varied from station to station, transportation was the principal community noise compo-

Acoustical Survey of Community Noise

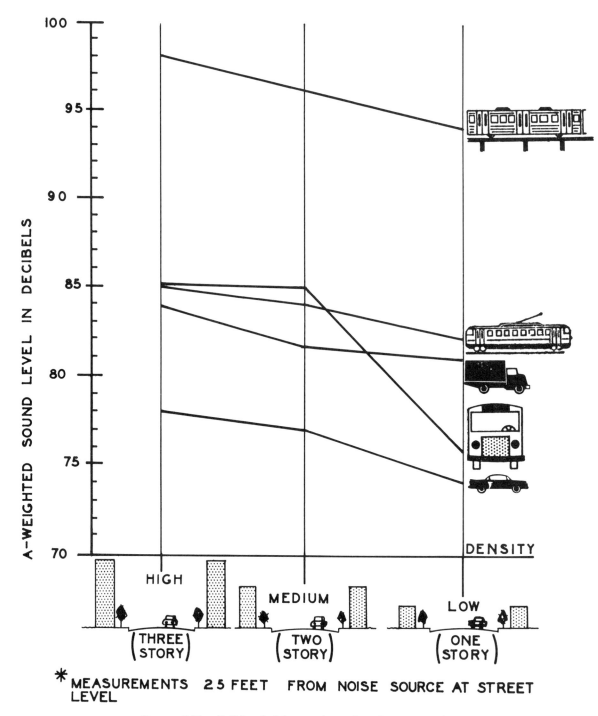

FIGURE 5.16. *Building height as a factor in noise propagation*

nent in both communities. Aircraft was the dominant noise source in Tinicum, because the Township lies more or less beneath the flight path. Automobiles and trucks were the predominant noise sources in Philadelphia (Stations 6, 7, 16, etc.). However, at those stations having public transportation, the noise was equal to or exceeded automobile and truck noise. At Stations 1 and 20, the trolley had the highest decibel level of any noise source, while at Station 2, it was the elevated-subway.

The sampling schedule was designed to evaluate the significance of time as a factor affecting community noise intensity. Are there periods of the day when the din of our cities is noticeably different? At least in Philadelphia, there were no appreciable changes in the noise level attributable to a particular time of day (between 7:00 A.M. and 11:00 P.M.).

TIME OF DAY

Hour by hour, readings at our monitoring stations resulted in only minor variation. For statistical purposes, the day was then arbitrarily divided into morning (7:00 A.M.–12:00 noon), afternoon (noon–6:00 P.M.), and evening (6:00 P.M.–11:00 P.M.) time periods, and noise levels were again compared (see Figures 5.17–5.19). These comparisons did not yield any major differences either. In Tinicum the morning, afternoon, and evening periods produced nearly identical noise levels, with the differences from morning to evening being one decibel (55 to 56 dBA). For West Philadelphia, the range of differences was three decibels (55 to 58 dBA).

However, some monitoring stations had a larger variation than others. Five stations recorded a variation of eight to ten decibels between morning and evening hours. However, at all the Tinicum locations and at nine Philadelphia stations, time was a negligible factor; the variation in noise never exceeded six decibels. The lowest noise level tended to be recorded during the evening period, as opposed to the morning or afternoon. Generally speaking, however, the evening hours offered the city inhabitants little relief from the day-long noise.

Only at night (11:00 P.M.–6:00 A.M.) did noise appreciably diminish. Based on the sampling schedule completed during these hours, the mean noise level in Philadelphia was 40 dBA. Among the specific stations, the range of intensity varied from 34 dBA to 46 dBA. *Average peak readings approached those recorded during the daylight hours, but the peaks were usually shorter in duration and fewer in occurrence.* Consequently the average minimum noise level dropped sharply, down to 37 dBA.

Mobility, which characterizes the degree of community activity, was also evaluated. How does increased activity during certain periods of

COMMUNITY MOBILITY

MONITORING STATIONS

WEST PHILADELPHIA (stations 1–20)   **TINICUM TWP.** (stations 1–5)   **NOISE LEVEL AVG. IN DBA**

TIME	1	2	3	4	5	6	7	8	9	10	11	12	13	14	15	16	17	18	19	20	1	2	3	4	5	TIN.	W.PHIL.	TOTAL
DAY AVERAGE	65	62	48	54	53	64	64	51	53	60	53	50	52	61	55	63	53	59	56	65	56	58	54	53	56	55	57	57
MORNING AVERAGE	67	64	50	53	54	66	62	55	54	59	55	52	55	57	58	62	54	57	56	63	55	58	55	54	53	55	57	57
AFTERNOON AVERAGE	65	64	51	53	54	66	64	62	55	62	56	51	55	62	58	66	54	60	58	64	58	58	55	53	57	55	55	58
EVENING AVERAGE	63	58	43	55	53	58	62	45	52	58	46	46	47	63	47	60	52	60	54	57	57	58	55	53	57	56	56	56
MOBILITY																												
HIGH MOBILITY	67	64	50	56	54	66	65	53	54	61	55	52	55	61	58	64	54	59	55	66	58	61	56	56	57	57	58	58
MEDIUM MOBILITY	63	61	47	52	53	62	63	49	53	59	51	48	51	62	53	62	52	59	57	64	55	55	53	50	56	53	56	55
WEEKDAY AVERAGE	65	63	49	54	54	66	64	51	54	60	54	52	55	61	57	64	53	60	55	65	56	58	55	53	56	55	56	56
WEEKEND AVERAGE	67	62	49	54	52	60	65	52	52	59	51	47	50	62	53	61	53	58	57	65	56	59	53	54	56	56	56	56

FIGURE 5.17. *Community noise level by monitoring station: mean*

MONITORING STATIONS

WEST PHILADELPHIA (stations 1–20)   **TINICUM TWP.** (stations 1–5)   **NOISE LEVEL AVG. IN DBA**

TIME	1	2	3	4	5	6	7	8	9	10	11	12	13	14	15	16	17	18	19	20	1	2	3	4	5	TIN.	W.PHIL.	TOTAL
DAY AVERAGE	88	89	72	78	72	84	83	72	75	86	78	71	77	84	78	85	77	86	77	87	87	82	82	86	77	82	79	81
MORNING AVERAGE	93	89	77	78	75	81	84	78	74	87	79	72	77	80	84	81	77	88	73	86	81	86	87	87	71	82	80	81
AFTERNOON AVERAGE	87	90	76	76	71	88	81	72	72	90	83	75	85	86	79	87	77	85	82	86	88	78	81	82	73	80	81	81
EVENING AVERAGE	84	90	62	80	72	83	81	66	82	83	69	69	70	86	70	90	78	85	74	88	87	85	81	90	84	85	79	81
MOBILITY																												
HIGH MOBILITY	90	88	79	80	72	84	83	76	74	90	82	75	83	82	83	84	80	88	76	89	89	87	82	91	79	85	81	83
MEDIUM MOBILITY	87	91	65	77	72	84	84	68	76	83	74	68	72	86	73	87	75	84	78	85	85	78	82	82	76	80	78	79
WEEKDAY AVERAGE	90	89	75	79	74	86	84	73	73	87	80	76	79	85	81	85	78	88	76	87	88	82	85	87	79	84	80	81
WEEKEND AVERAGE	86	90	69	75	66	81	83	75	73	87	76	65	77	81	76	86	75	79	79	88	83	83	82	86	73	80	78	79

FIGURE 5.18. *Community noise level by monitoring station: peak*

	MONITORING STATIONS — WEST PHILADELPHIA																				TINICUM TWP.					NOISE LEVEL AVG. IN DBA		
TIME																												
DAY AVERAGE	50	49	43	45	47	50	48	43	44	47	46	42	45	50	44	48	46	49	46	51	45	47	46	44	46	45	46	47
MORNING AVERAGE	49	51	45	46	46	53	49	45	46	49	47	45	46	48	45	48	47	48	47	50	45	47	46	46	46	46	47	47
AFTERNOON AVERAGE	54	51	43	46	48	52	48	43	45	49	47	43	46	53	43	50	46	51	48	51	40	46	48	45	46	47	45	47
EVENING AVERAGE	48	44	41	46	46	43	45	41	41	42	43	40	42	50	43	44	45	49	44	50	45	49	46	42	46	45	46	46
MOBILITY																												
HIGH MOBILITY	52	51	44	46	48	52	49	44	45	48	46	43	46	51	44	48	48	49	46	53	46	49	48	45	48	47	47	48
MEDIUM MOBILITY	48	48	43	45	46	48	48	43	43	46	46	42	44	50	45	48	44	49	47	49	44	46	45	43	44	44	46	46
WEEKDAY AVERAGE	50	50	43	46	47	52	49	43	46	47	46	43	46	51	45	49	46	50	45	51	45	49	48	44	47	46	47	47
WEEKEND AVERAGE	51	49	44	46	46	45	46	45	44	50	45	47	43	50	44	46	45	48	48	49	45	44	45	44	45	44	46	46
MONITORING STATION NUMBER	1	2	3	4	5	6	7	8	9	10	11	12	13	14	15	16	17	18	19	20	1	2	3	4	5	TIN.	W.PHIL.	TOTAL

FIGURE 5.19. *Community noise level by monitoring station: low*

the day affect the level of community noise? How significant is this effect?

The noise levels recorded during periods of high and medium mobility were very similar (see Figures 5.17–5.19). During high mobility (7 A.M.–10 A.M., and 3 P.M.–7 P.M.), the mean noise level was 3 dBA greater than during medium mobility (10 A.M.–2 P.M., and 7 P.M.–11 P.M.), or 58 dBA as opposed to 55 dBA. In Tinicum the difference was 4 dBA (57 dBA versus 53 dBA), while in Philadelphia it was 2 dBA (58 dBA versus 56 dBA). There was no individual station where the range between high and medium exceeded 6 dBA.

Only during periods of low mobility (11 P.M.–6 A.M.) was a sizable change in the noise level observed. Although data covering this period are incomplete, the mean noise level in Philadelphia was 40 dBA, or 16 dBA less than during medium mobility.

Aircraft Noise

According to the field survey, aircraft causes the most pervasive noise problem. It particularly affects those living in Tinicum Township. Here the noise level of landing aircraft rises as planes approach the east-west runway, located just beyond a populated section of Tinicum. As planes fly over the Township from the west, their altitude on approach decreases (using a three-degree glide slope angle) approximately 400 feet between Taylor and 4th Avenues (see Figure 5.14). Accompanying this drop in altitude there is a rise in the noise level of nine to ten decibels, depending upon the type of aircraft (see Figure 5.2). For example, a three-engine jet on approach registers 93 dBA at 2.5 miles from the runway (where the flight path crosses Taylor Avenue). By the time such a plane is 1.25 miles from the runway, where it crosses 4th Avenue, the noise level has risen by ten decibels, to 103 dBA. Beginning at Taylor Avenue, for every 100-foot descent along the flight path, flyover noise increases roughly 2.5 decibels, (assuming a three-degree glide slope angle).

The most intense noise levels were measured directly below the flight path, with the peak level occurring at the intersection of 4th Avenue and Second Street (see Figure 5.2 and 5.3). The noise levels of landing aircraft measured at various locations within the Township are presented in Table 5.6. These levels represent the average readings recorded for piston and turboprop aircraft, and turbofan and turbojet aircraft. Besides the five monitoring stations three other locations were used for measuring the sound pressure level of overflights. Seven of these eight locations recorded jet noise in the mid 80 dBA range or above. It is approximately at this level that noise begins to be annoying on a community wide basis. Somewhere between 82 dBA and 92 dBA there is a threshold of human tolerance to aircraft noise.[6] Above this threshold even those normally favorably disposed become intolerant. The only area where aircraft do not appear to be annoying while landing is the northern part of Lester, (see Figure 4.2 for geographical reference) north of Governor Printz Boulevard in the vicinity of Station 5.

TABLE 5.6.   Mean Sound Pressure Level of Landing Aircraft: Tinicum Township

Geographic location	Piston & turboprop aircraft		Turbofan & turbo-jet aircraft	
	dBA	PNdB	dBA	PNdB
Station One: 4th & Seminole Avenues	72–76	85–89	86–90	99–103
Station Two: 3rd & Wanamaker Avenues	75–79	88–92	86–93	99–106
Station Three: 3rd & Saude Avenues	73–78	86–91	84–90	97–103
Station Four: 2nd & Jansen Avenues	86–89	99–102	96–103	109–116
Station Five: 3rd & Seneca Avenues	64	77	68–70	81–83
Putcan & Front Streets	86–90	99–103	97–102	110–115
Taylor Avenue & Governor Printz Park	92–94	105–107	100–108	113–121
4th Avenue & 2nd Street	80–84	93–97	89–97	102–110

Mean Sound Pressure Level of Aircraft in Takeoff: Tinicum Township   TABLE 5.7.

Geographic location	Piston & turboprop aircraft		Turbofan & turbo-jet aircraft	
	dBA	PNdB	dBA	PNdB
Station One: 4th & Seminole Avenues	—	—	89–97	102–110
Station Two: 3rd & Wanamaker Avenues	75–79	88–92	80–92	93–105
Station Three: 3rd & Saude Avenues	—	—	86–94	99–107
Station Four: 2nd & Jansen Avenues	—	—	87–94	100–107
Station Five: 3rd & Seneca Avenues	70–82	83–95	82–97	95–110
Putcan & Front Streets	—	—	84–102	97–115
4th Avenue & 2nd Street	76–80	89–93	86–100	99–113

Varying according to the pattern and destination of flight, departing aircraft during a day's time flew over all populated areas of Tinicum. No part of the Township was continually protected from the noise of aircraft during takeoff. In comparison to landing, the noise during takeoff was more evenly distributed (see Table 5.7). All the measuring points had an average noise level for turbofan- and turbojet-powered air carriers above 80 dBA. Although the highest levels in takeoff were below those for landing aircraft, a greater portion of the community was annoyed by takeoff due to the wider distribution of takeoff patterns. These flights were consistently around the threshold of human tolerance to noise.

Mean noise levels were computed by type and size of aircraft. Within each category there was considerable variation in noise intensity. A four-engine jet, while landing, averaged 103 dBA at Station 4, but noise measurements ranged from 94 to 111 dBA, or 107 to 124 PNdB (see Table 5.8). Often there was overlapping, so that the noisiest two-engine jet had a higher sound level than the quietest three-engine jet, and so forth.

These measurements taken at ground level indicate that *most of the Tinicum residents, and especially those in the flight approach path, are exposed to intense aircraft noise. Here the sound pressure levels approach and frequently surpass those recorded in any known residential community near an airport.* Along the approach path, beginning at Lagrange Street, four-engine jet planes usually exceeded 112 PNdB. This is the maximum allowable noise level for aircraft established at the John F. Kennedy

TABLE 5.8.   Sound Pressure Level of Landing Aircraft at Station 4:
Tinicum Township

Aircraft		Sound level						Sample
Piston &	Turbofan &	Low		Peak		Average		size
turboprop	turbojet	dBA	PNdB	dBA	PNdB	dBA	PNdB	
2 engine		80	93	90	103	86	99	12
4 engine		87	100	91	104	89	102	3
	2 engine	88	101	100	113	94	107	17
	3 engine	94	107	103	116	97	110	36
	4 engine	94	107	111	124	103	116	33

International Airport in New York, and also at London's Heathrow
Airport.[7]

EXISTING AIRCRAFT
OPERATIONS

Airplane movements have been steadily increasing since the
Philadelphia International Airport opened. These movements constitute
the total number of aircraft arrivals and departures for a prescribed
period of time. They are also referred to as aircraft operations.

For statistical purposes aircraft operations are classified according
to four categories: air carrier, military itinerant and local, civil itinerant,
and civil local. Air carrier flights, both commercial passenger and cargo,
represent the largest category (see Table 5.9). Of the nearly 300,000
total plane movements in 1968, 197,000, or 67 percent, were air carrier
movements.[8] Second were civil itinerant aircraft movements (small-
engined, privately operated planes) with 30 percent of the total. Military
itinerant and local, and civil local composed the remaining 3 percent,
or 4,900 movements in 1968. Air carrier type operations are the most
important category. They are the largest source of aircraft noise, and
they represent the bulk of all flights using Philadelphia International
Airport.

Distribution of Air
Carrier Movements

During 1968 there were, on the average, 540 daily air carrier movements. Each year the average daily rate rises; compared to 1967 this
represents an increase of 78 movements per day, and 161 more movements per day than in 1966. Compared to other forms of transportation,
the number of aircraft operations varies considerably, depending upon
the season, month, day, and even hour.

For the year 1968, the average number of monthly operations was
16,445, but certain months differed significantly from this average (see
Figure 5.20).[9] August was the peak month, February the lowest. With
the exception of December, air travel is at its lowest point during the
coldest months (November–February). During the warmer months,
air travel grows. Coincident with the air traffic peak, outside residential

TABLE 5.9.

Aviation Activity Report: Philadelphia International Airport: 1963–1968

ACTIVITY	1963	1964	1965	1966	1967	1968
1. *Plane movements*						
(a) Air carrier	123,684	128,985	135,324	138,358	168,908	197,345
(b) Military itinerant & local	6,516	5,638	4,638	4,975	4,073	3,611
(c) Civil itinerant	61,990	67,093	74,628	85,376	85,317	90,940
(d) Civil local	5,276	4,322	3,325	2,542	1,919	1,299
TOTAL PLANE MOVEMENTS	197,466	206,038	217,915	231,251	260,217	293,195
2. *Passenger traffic*						
(a) Scheduled domestic						
(1) Arrivals	1,362,013	1,557,757	1,792,534	2,026,680	2,474,581	3,059,619
(2) Departures	1,346,976	1,511,000	1,795,622	1,995,081	2,482,816	3,063,747
(b) Non-scheduled, charter & itinerant						
(1) Arrivals	18,435	14,791	22,895	33,492	37,194	42,299
(2) Departures	10,278	14,421	17,333	20,934	28,456	48,446
(c) Scheduled international						
(1) Arrivals	18,459	21,149	27,798	35,449	60,198	68,806
(2) Departures	22,821	20,070	29,798	34,533	55,280	67,771
(d) Charter & itinerant international						
(1) Arrivals	5,867	5,813	5,633	7,827	10,395	23,575
(2) Departures	4,809	4,611	3,429	5,624	7,510	19,010
TOTAL PASSENGER TRAFFIC	2,789,658	3,149,612	3,695,042	4,159,620	5,156,430	6,393,273
3. *Air Mail (tons)*						
(a) Domestic	6,967	7,807	9,715	12,664	19,384	26,371
(b) International	22	4	N/A	6	26	28
TOTAL AIR MAIL	6,989	7,811	9,715	12,670	19,410	26,399
4. *Air freight traffic (tons)*						
(a) Scheduled domestic	29,819	39,280	54,862	64,553	76,034	93,702
(b) Non-scheduled domestic	6,367	6,586	9,739	9,008	7,631	5,382
(c) Scheduled international	464	191	744	1,355	1,802	2,747
(d) Non-scheduled international	130	244	109	—	—	33
TOTAL AIR FREIGHT	36,780	46,301	65,454	74,916	85,467	101,864
5. *Air Express (tons)* TOTAL	6,637	7,413	8,769	8,370	7,820	8,448

Source: "Aviation Activity Report: Philadelphia International Airport," City of Philadelphia, Division of Aviation, Administrative Branch, March 20, 1969.

142

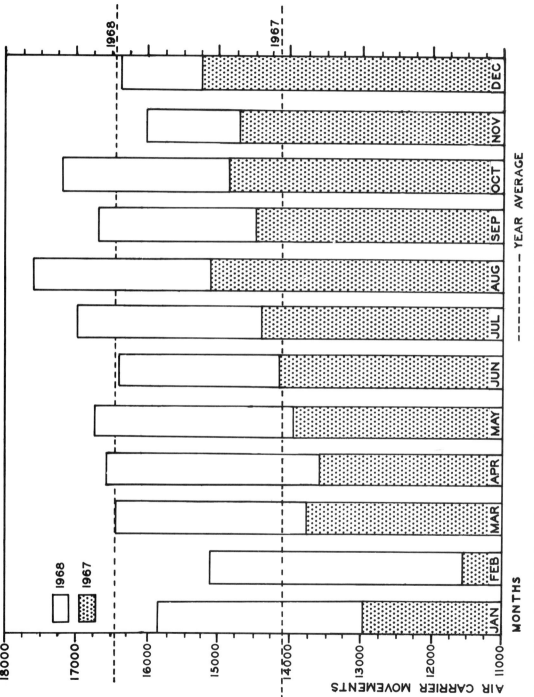

FIGURE 5.20. *Scheduled air carrier movements by month: Philadelphia Airport, 1967 and 1968. Data from "Aircraft Operations: 1967–1968," City of Philadelphia, Division of Aviation, Administrative Branch.*

activity is at its height. This is the time when community residents are likely to be most annoyed by aircraft noise.

On Saturdays and Sundays, there are slightly fewer air carrier movements than on weekdays. In September 1967 there were 248 arrivals per day during the weekdays and 236 on Saturdays, 237 on Sundays. Movements vary little during weekdays at Philadelphia Airport. The same is true of air freight flights, except on Mondays, when there are no scheduled arrivals or departures. However, the pattern changes on weekends. Air freight traffic is reduced by 50 percent on Saturdays, and none is scheduled on Sundays.

The distribution of flights over the day indicates certain preferred travel times. Late afternoon, between 4:00 and 6:00 P.M., are the peak hours (see Figure 5.21).[10] Nearly 10 percent of all daily aircraft movements occur at this time. During the morning hours from 8:00 A.M. to 10:00 A.M. there is a secondary peak. In the evening the largest hourly total of scheduled aircraft movements takes place between 9:00 and 10:00 P.M. Although the airport traffic falls off considerably after midnight, there is at least one scheduled flight every hour of the night; between 12:00 and 7:00 A.M. there are 54 scheduled flights. Air traffic during these hours, though still relatively small, has nearly tripled since 1965. Furthermore, since aircraft, most of them jet-powered, fly throughout the night, community residents do have their sleep disturbed.

Presently there are four operating runways. The principal one, Runway 9, runs east to west (see Figure 4.2). No statistical records are maintained on runway usage, but the airport authorities estimate that Runway 9 handles approximately 60 to 70 percent of all aircraft operations.[11] If only air carrier movements are considered, that percentage increases to 80 to 90 percent, since only this runway is capable of handling commercial jets.

Most of the time, air carriers approach Runway 9 from the west and take off to the east because of prevailing winds. However, wind patterns often change, particularly in the fall, and directions on approach and takeoff are then reversed.

Jet service into Philadelphia began in 1959. Since then, there has been a rapid decline in the use of propeller-driven air carriers. An engine survey of all flights arriving in Philadelphia International during September of 1967 indicated that only 4 percent of the arriving aircraft were powered by reciprocating engines; the remaining 96 percent were jet-powered.[12] Nearly two thirds, 61 percent, of all commercial aircraft had pure-jet or turbofan engines; the balance, 35 percent, had turboprop engines. As the demand for greater payloads and larger aircraft continues, the noise of the turbofan engine grows. Efforts to

Runway Usage

Aircraft Mix

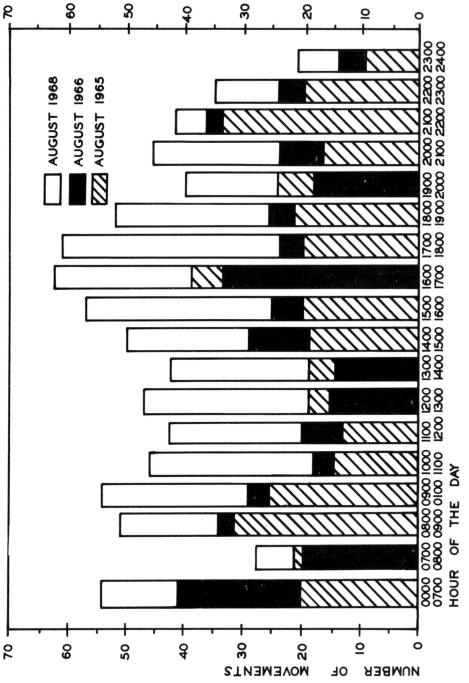

FIGURE 5.21. *Scheduled aircraft movements: Philadelphia International Airport. Data from City of Philadelphia, Division of Aviation, Administrative Branch.*

control engine noise have up to this point resulted in only modest reductions. Newer generation aircraft engines, though slightly less intense, still produce noise levels annoying to community residents.

The number of aircraft movements has grown rapidly. In 1967 the *percentage* of growth, with an increase of 18 percent over 1966, exceeded any previous year (see Table 5.9). The growth *rate* does not appear to be slowing either. The increase in air carrier movements for 1968, with another 28,437 movements, nearly equalled the 1967 record increase. In two years, 1967 and 1968, the daily number of arrivals and departures grew by 170.

In early 1968 Philadelphia International surpassed the Federal Aviation Agency projection for it of 171,000 air carrier movements for 1970 (see Figure 5.22).[13] Even though the estimate is conservative, based on the FAA forecast, by 1975 air carrier operations will have grown another 30, and by 1980 another 40 percent. *In ten short years air carrier operations can be expected nearly to double.* The noise problem will become even more severe.

During certain periods of the day, the airport is even now operating to capacity. Most future growth in air traffic will therefore occur at off-peak hours. Passenger flights will increase, but not as rapidly as air freight flights. Passenger aircraft used during the day and early evening hours are converted to handle air freight during later evening and nighttime hours. This allows fuller utilization of aircraft and therefore is economically attractive to airlines and airport operators. *With the greater use of aircraft, airport operations will extend further into the night, and the periods of quiet will become shorter.*

A parallel east-west runway is now under construction. Airport traffic will be somewhat altered, and with it the noise it generates (see below).

In early 1968 construction began on an east-west runway parallel to the existing one (see Figures 5.23 and 5.24). This new 1050-foot runway is being built at a cost of over $16,000,000 and is slated to be open by the end of 1971 or early 1972. When opened, it will nearly double the capacity of commercial jet operations. Essentially all jet aircraft are now restricted to Runway 9; therefore, landings and takeoffs cannot occur simultaneously.

Noise level contours for jet aircraft (both landing and taking off) were superimposed onto Township base maps to compare the existing runway with the one being completed (see Figures 5.23 and 5.24). The source of these contours, as mentioned earlier, was work done by Bolt, Beranek, and Newman, for predicting noise levels at any given airport facility.[14]

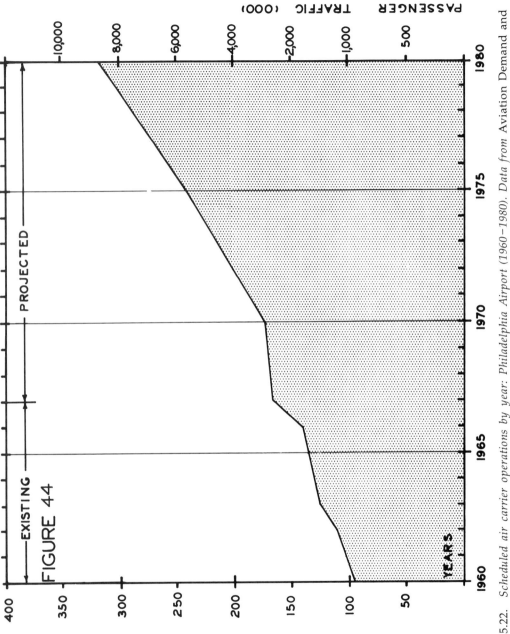

FIGURE 5.22. *Scheduled air carrier operations by year: Philadelphia Airport (1960–1980). Data from Aviation Demand and Airport Facility Requirements: Forecasts for Large Air Transportation Hubs through 1980, Federal Aviation Authority, Department of Transportation, 1967.*

FIGURE 5.23. *Perceived noise level contours of landing turbofan and turbojet aircraft: Tinicum Township*

TABLE 5.10.   Noise Reduction within Residences: Tinicum Township Station 4

Jet aircraft flyover noise	House I	House II	Average
Mean level outside	100 dBA	94 dBA	97 dBA
Mean level inside	80 dBA	65 dBA	72 dBA
Mean difference	20 dBA	29 dBA	25 dBA
Peak level inside	84 dBA	71 dBA	78 dBA
Lowest level inside	73 dBA	60 dBA	67 dBA

path and at each house 18 flyovers were recorded. One sound level meter was placed in a bedroom while the other was placed outside, ten feet from the house and four feet above the ground. At the time of the survey both of these single-family detached residences had storm windows in place with all doors and windows closed. The principal differences between the two were type of construction (House I: brick; House II: wood frame) and the interior microphone location (House I: second floor bedroom; House II: first floor bedroom).

At House I, the mean difference between jet aircraft noise measured inside and outside was 20 dBA, while the difference was 29 dBA at House II (see Table 5.10). A major reason for the variation between these houses was the location of the interior microphone. Second-floor readings would be expected to be higher than first-floor readings. In both residences the mean noise level was extremely high, particularly for sleeping (House I: 80 dBA; House II: 65 dBA). At no time during this survey, 8:00 P.M.–11:00 P.M., did the level of aircraft noise inside these homes register below 60 dBA.

It would be expected that during warmer weather when storm windows are removed and houses are "opened up," noise intrusion would be even a bigger problem. The interior mean noise level for the two residences would rise from 72 dBA to 80–85 dBA. When windows are even partially opened, the amount of noise reduction within a dwelling decreases to approximately 10 decibels.[17]

VI

The Metropolitan Philadelphia Noise Survey: Analysis of Community Response

At the beginning of our survey, a series of general questions was presented to residents of Tinicum and West Philadelphia (see Appendix A) to appraise the population's attitude toward their local environment. They were asked, among other things, how satisfied they were with the area in which they lived, which of its features they liked best and which least, and how they would rank a series of environmental problems. *At this time no questions were asked dealing directly with noise;* any mention of noise in the answers is therefore wholly spontaneous. An additional series of questions was intended to determine the residents' activity patterns, including the extent of their mobility.

Residential Environment

Approximately two thirds of those queried judged their area to be either an excellent (12%) or a good (52%) place to live generally, while a smaller group (7%) gave it a poor rating (see Table 6.1). Comparing the Tinicum and West Philadelphia responses, the Tinicum residents were more satisfied with their environs than the West Philadelphia residents (21% in Tinicum rated their area as excellent as opposed to only 10% in West Philadelphia).

DEGREE OF RESIDENTIAL SATISFACTION

TABLE 6.1.   Residential Satisfaction

Residential rating	Tinicum Township		West Philadelphia		Total	
	Number	%	Number	%	Number	%
Excellent	21	21.0	39	9.8	60	12.0
Good	53	53.0	210	52.9	263	52.6
Fair	20	20.0	122	30.0	142	28.4
Poor	6	6.0	29	7.3	35	7.0
TOTAL	100	100.0	400	100.0	500	100.0

Noise was found to be a major factor influencing people's judgment of their residential environment. A chi-square analysis indicates a significant interaction (P<.01) between the variable of residential satisfaction and the perceived noise intensity (see Table 6.2). The data show that those most satisfied with the area perceive it as quiet, while those considering their area a noisy place to live tend to be less satisfied with it.

POSITIVE CHARACTERISTICS OF THE NEIGHBORHOOD

"Neighborliness" was the most frequent answer to the question "What do you like best about the area in which you live?" (see Table 6.3). Convenience of transportation was second, shopping convenience third. If all aspects of convenience (transportation, shopping, employment, and general) are combined, they constitute the largest response category, exceeding neighborliness by 344 to 131. The people surveyed in Philadelphia mentioned convenience more often than the Tinicum Township people, who have fewer local shopping facilities and no public transportation. After convenience and neighborliness came the

TABLE 6.2.   Residential Satisfaction and Perceived Noise Intensity

Rating of neighborhood	Perceived noise in neighborhood					Total
	Very quiet	Fairly quiet	Slightly noisy	Noisy	Very noisy	
Excellent	19	24	7	3	7	60
Good	44	111	64	28	14	261
Fair	6	46	38	38	16	134
Poor	4	7	5	8	10	34
TOTAL	73	188	114	77	47	489

Chi square 73.11017, Level of significance P<.001
The chi-square analysis on this table and in the following tables is computed on all columns.

Perceived Positive Residential Characteristics   TABLE 6.3.

Category	Tinicum Township Number	%	West Philadelphia Number	%	Total Number	%
Neighborliness	28	28.00	103	25.75	131	26.20
Convenience: Shop	1	1.00	102	25.55	103	10.60
Convenience: Work	16	16.00	33	8.25	49	9.80
Convenience: Transp.	4	4.00	114	28.50	118	23.60
Home	5	5.00	48	12.00	53	10.60
Privacy	4	4.00	10	2.50	14	2.80
Safety	2	2.00	9	2.25	11	2.20
Schools	11	11.00	39	9.75	50	10.00
Church	0	0.00	17	4.25	17	3.40
Recreation	8	8.00	13	3.00	21	4.20
Other Municip.– Service	5	5.00	12	3.00	17	3.40
Tax	23	23.00	1	.25	24	4.80
Native	23	23.00	19	4.75	42	8.40
Open Space	15	15.00	40	10.00	55	11.00
Quiet (excluding planes)	11	11.00	48	12.00	59	11.80
General Convenience	9	9.00	65	16.25	74	14.80
Nothing	9	9.00	30	7.50	39	7.80
Other	23	23.00	46	11.50	63	12.60
SAMPLE SIZE	100	–	400	–	500	–

amenity of quiet. However, many prefaced their answer of "quiet" by saying *"except for aircraft noise."* Nearly 8 percent did not believe their area had any particularly positive qualities. The same 8 percent also felt their neighborhood to be noisy. These answers are not, however, attributable to the population of any one particular monitoring station.

To the question, "What do you like least about the area where you live?" more people gave the answer "nothing" than any other (see Table 6.4). The majority of those responding in this way also felt their neighborhood was a quiet place in which to live. Noise represented the leading source of expressed residential dissatisfaction, numbering 186 out of a total of 735 responses. The population objected to aircraft noise more often than to any other type of noise. Nearly all of the aircraft noise complaints came from Tinicum residents. After aircraft noise, they most disliked the noise of vehicles and children. Other expressed dislikes, besides noise, concerned the condition of the neighborhood streets, crime, air pollution, municipal services, and shopping and public transportation facilities. A sizable percentage of the Philadelphia

NEGATIVE
CHARACTERISTICS
OF THE
NEIGHBORHOOD

TABLE 6.4.   Perceived Negative Residential Characteristics

Category	Tinicum Township Number	%	West Philadelphia Number	%	Total Number	%
Crime	4	4.00	53	13.25	57	11.40
Air pollution	8	8.00	23	5.75	31	6.20
Noise: Aircraft	55	55.00	5	1.25	60	12.00
Vehicles	14	14.00	19	4.75	33	6.60
Subway-St. cars	1	1.00	14	3.50	15	3.00
Trains	3	3.00	12	3.00	15	3.00
Children	1	1.00	30	7.50	31	6.20
General	8	8.00	24	6.00	32	6.40
Neighborhood condition:						
Housing	0	0.00	20	5.00	20	4.00
Streets	0	0.00	59	14.75	59	11.80
Industry	9	9.00	15	3.75	24	4.80
Inconvenience:						
Shopping	14	14.00	31	7.75	45	9.00
Transportation	20	20.00	11	2.75	31	6.20
Municipal services	1	1.00	33	8.25	34	6.80
Neighborhood changing	1	1.00	47	11.75	48	9.60
Nothing	7	7.00	101	25.25	108	21.60
Other	22	22.00	70	17.50	92	18.40
SAMPLE SIZE	100	—	400	—	500	—

residents (11%) also expressed concern over the racial change of their neighborhood, either from black to white or from white to black.

AWARENESS OF LOCAL ENVIRONMENTAL PROBLEMS

Each respondent was asked, "In your area, to what degree do you have problems with air pollution, crime, abandoned vehicles, poor housing, littering, and noise?" The answers were compared and ranked according to the number of responses indicating that some problem existed (see Table 6.5). When, in contrast with earlier open-ended ques-

TABLE 6.5.   A Rank Order of Urban Environmental Problems

Environmental problems	Tinicum Township Number	%	West Philadelphia Number	%	Total Number	%
Air pollution	82	82.0	293	73.3	375	76.8
Littering	49	49.0	268	67.0	317	65.0
Noise	88	88.0	236	59.0	317	65.0
Crime	40	40.0	275	69.0	315	64.0
Abandoned vehicles	20	20.0	166	41.0	186	38.3
Poor housing	18	18.0	168	42.0	186	38.3

Awareness of Environmental Problems and Perception of Noise
Intensity: Chi Square Analysis

TABLE 6.6.

Environmental problems	Tinicum Township Level of significance	West Philadelphia Level of significance	Total Level of significance
Air pollution	.10 (trend)	.10 (trend)	.05
Crime: Juvenile delinquency	−	.001	.001
Abandoned vehicles	−	.01	.10 (trend)
Poor housing	−	.001	.10 (trend)
Littering	−	.001	.001
Noise	.001	.001	.001

tions, specific problems were listed, noise ranked behind air pollution
and littering but above crime, abandoned vehicles, and poor housing.
Nearly two thirds of the respondents mentioned noise as a problem.
This figure represents 88 percent of the Tinicum respondents and 59
percent of those interviewed in West Philadelphia. Another opinion
survey conducted in Philadelphia by Ring and Schein found similar
results.[1] Noise, it appears, is one of the most commonly recognized
problems in both areas of our survey.

A statistical analysis by chi square correlating awareness of environmental problems and perception of noise intensity appears highly
significant (see Table 6.6) in nearly every instance. People believing
noise to be a problem generally also felt there were other environmental problems.

The type of noise most annoying to the survey population was
that of aircraft (see Table 6.7). In Tinicum this response was nearly
universal (87 percent). As expected, the proportion of those annoyed
outside Tinicum was much less, registering a mere 13 percent; the
Philadelphia monitoring stations closest to the airport accounted for
most of these. The second leading annoyance, commonly mentioned by
city residents only, was noise made by children.

Industrial noise appeared to be more of a problem in Tinicum than
in West Philadelphia. Residents in both areas considered vehicular
traffic annoying. Truck traffic was a bigger problem to the Tinicum
people than automobile traffic, since most of Tinicum is removed from
through-traffic. Activities of a transitory nature, specifically building
and road construction noises, were mentioned infrequently. Noises
of this type were usually tolerated more readily because inhabitants
seemed to feel the condition was only temporary. It is interesting that

*Personal Reactions
to Noise*

GENERAL
ANNOYANCE

Noise Sources

TABLE 6.7.   Annoyance by Type of Noise

Response	Tinicum Township Number	%	West Philadelphia Number	%	Total Number	%
Aircraft	87	87.00	52	13.00	139	27.80
Public transportation	None	–	44	11.00	44	8.80
Vehicle traffic: General	7	7.00	64	16.00	71	14.20
Vehicle traffic: Truck	15	15.00	45	11.25	60	12.00
Industry	11	11.00	12	3.00	23	4.60
Construction	1	1.00	6	1.50	7	1.40
Children	4	4.00	114	28.50	117	23.60
Adults	0	0.00	28	7.00	28	5.60
Animals	5	5.00	24	6.00	29	5.80
Railroad	6	6.00	23	5.75	29	5.80
Other	6	6.00	58	14.50	64	12.80
Not annoyed	4	4.00	108	27.00	112	22.40
No answer	0	0.00	24	6.00	24	4.80
Number in sample	100	–	400	–	500	–

in Tinicum nearly no annoyance at all was expressed concerning noise made by children or adults. In Philadelphia, where the residential density is higher but the household size similar, well over 35 percent of the respondents were bothered by noise of this type.

Time of Day or Week

More people felt that disturbing noise occurred continuously throughout the day rather than only during specific time periods (see Table 6.8). Among those specifying a particular time of day, objections were strongest to noises occurring during the night or evening hours.

TABLE 6.8.   Noise Occurrence and Perceived Annoyance: Time of Day

Time	Tinicum Township Number	%	West Philadelphia Number	%	Total Number	%
Morning	0	00.00	24	6.33	24	5.10
Afternoon	8	8.70	24	6.33	32	6.79
Evening	32	34.78	45	11.87	77	16.35
Night	20	21.74	73	19.26	93	19.75
All the time	19	20.65	119	31.40	138	29.30
None of the time	0	00.00	27	7.13	27	5.73
Morning & afternoon	6	6.52	27	7.13	33	7.01
Afternoon & evening	3	3.26	34	8.97	37	7.86
Morning & night	4	4.35	6	1.58	10	2.11
TOTAL	92	100.00	379	100.00	471	100.00

Noise Occurrence and Perceived Annoyance: Day TABLE 6.9.

Day	Tinicum Township		West Philadelphia		Total	
	Number	%	Number	%	Number	%
Weekdays	7	10.15	87	47.03	94	37.01
Weekends	33	47.83	51	27.57	84	33.07
Both	29	42.03	47	25.41	76	29.92
TOTAL	69	100.00	185	100.00	254	100.00

Tinicum residents appeared to be annoyed more often by noise during the evening than noise at night, with aircraft representing the primary source of this annoyance. Very few in Tinicum indicated they were never annoyed by some type of community noise.

The Township population tended to be bothered by aircraft either continuously or only on weekends; seldom weekdays only. This differed considerably from the Philadelphia sample (see Table 6.9). In the city, the distribution was more equal. Approximately one half of those disturbed reported being annoyed during weekdays, while the remainder indicated either weekends or the entire week. Approximately one person in three commented that he was annoyed by noise throughout the entire week.

Most people were annoyed by noises having an irregular pattern (see Table 6.10). Random noise makes human adaptation more difficult (see Chapter III). One exception was the noise associated with children, which to many respondents in West Philadelphia was annoying even though predictable.

Pattern of Noise

The location of a source of noise in relation to the person does not affect his response. In our survey, only a small number indicated that they were more annoyed by outside noise than by noise when inside the house (see Table 6.11).

Location of Noise

Noise Pattern and Perceived Annoyance TABLE 6.10.

Response	Tinicum Township		West Philadelphia		Total	
	Number	%	Number	%	Number	%
Irregular	85	93.41	259	68.88	344	73.66
Regular	6	6.59	117	31.12	123	26.34
TOTAL	91	100.00	376	100.00	467	100.00

TABLE 6.11. Location and Perceived Annoyance

Person's Location	Tinicum Township		West Philadelphia		Total	
	Number	%	Number	%	Number	%
Outside House	45	50.56	162	51.10	207	50.99
Inside House	33	37.08	137	43.22	170	41.87
Both	11	12.36	18	5.68	29	7.14
TOTAL	89	100.00	317	100.00	406	100.00

Activity Disruption

Our survey found few activities during which noise did not cause annoyance. Noise was considered most objectionable when it disrupted television (see Table 6.12). In Tinicum aircraft often masked the sound of TV and simultaneously disrupted the reception. Outside Tinicum, an additional 44 respondents mentioned their annoyance at interrupted television, but in their case ground transportation was the source. The second leading problem was interference with conversation. In Tinicum Township, residents refrained from talking when aircraft flew over, especially at Monitoring Stations 1–4. Talking on the telephone at such times was described as very difficult. Contributing to the annoyance was the fact that long distance telephone conversations were frequently interrupted, and these interruptions added to the duration and thus the cost of the calls. Almost equally disconcerting was noise interfering with sleep. Among Philadelphia residents it constituted the largest single complaint. The Township population ranked sleep interference third. There is a growing opinion that sleep disruption caused by community noise (discussed in Chapter III) may

TABLE 6.12. Degree of Annoyance during Selected Activities

Activity	Tinicum Township			West Philadelphia			Total		
	Often	Some-times	Never	Often	Some-times	Never	Often	Some-times	Never
Eating	28	68	1	12	33	329	40	101	230
Sleeping	49	47	1	46	110	220	95	157	221
Working	26	71	1	13	38	322	39	109	323
Reading	35	61	1	17	73	285	52	134	286
Television watching	88	9	1	34	90	250	122	99	251
Talking	82	15	1	20	75	281	102	90	282
Recreation	23	68	1	12	48	307	35	116	308
Relaxing	50	47	1	31	90	250	81	137	251

well be a greater problem than communication interference. In our study, several locations near major transportation systems were found to have noise levels sufficiently intense to affect the quality of sleep.[2] Many of the abatement techniques cited in Table 6.20 below were reported to be used at night to aid sleeping. The more routine activities of eating and working were least affected by community noises. In direct contrast, most of the pleasurable leisure time activities were liable to disturbance.

A subtle effect of noise previously unreported in community noise literature is the effect on interpersonal relations, particularly friendship. Those living in the airport community of Tinicum felt that the intense aircraft noise had affected their relations with others. They often commented that guests from other areas were noticeably uncomfortable visiting there. Instances were cited where close friends and even relatives declined to visit, particularly overnight, because they could not adjust themselves to the noisier environment. Outdoor entertaining during the warmer months around the airport was considered impossible. It seems likely that a worsening of the noise situation could significantly alter the social patterns of the Tinicum residents.

Spring appears to be the time of year when noise is most disturbing (see Table 6.13). The comparative stillness of winter passes with the advent of warmer weather. People become oriented again to outside activities, storm windows are removed, and community noise becomes more intrusive. It should not be assumed that the spring months represent the only time when people are disturbed by noise, but rather the time when it disturbs them most. Around Tinicum people are especially disturbed as long as the climate is moderate, from spring to early fall.

Seasonal Variation

Seasonal Variation and Perceived Annoyance

TABLE 6.13.

Response	Tinicum Township		West Philadelphia		Total	
	Number	%	Number	%	Number	%
No Answer	8	8.00	2	0.50	10	2.00
Winter	2	2.00	7	1.75	9	1.80
Spring	72	72.00	202	50.50	274	54.80
Summer	3	3.00	6	1.50	9	1.80
Fall	2	2.00	0	0.00	2	0.40
Spring–Summer	4	4.00	16	4.00	20	4.00
All	2	2.00	4	1.00	6	1.20
None	7	7.00	163	40.75	170	34.00
TOTAL	100	100.00	400	100.00	500	100.00

TABLE 6.14.   Response: Residence Affected by Noise?

Category	Tinicum Township		West Philadelphia		Total	
	Number	%	Number	%	Number	%
No answer	3	3.0	4	1.00	7	1.40
Yes	52	52.0	37	9.25	89	17.80
No	38	38.0	318	79.50	356	71.20
Possibly	7	7.0	41	10.25	48	9.60
TOTAL	100	100.0	400	100.00	500	100.00

AFFECT ON PROPERTY

Citizens often express concern about community noise because they believe it is affecting their property. Approximately 28 percent of the people believed their property either had been or could have been affected by noise (see Table 6.14). In Tinicum, this was the majority opinion. Of those who believed there was damage, the noise-related effects included broken windows, structural damage such as cracks in walls or ceilings, loss in property, and vibration (see Table 6.15). At one township monitoring station (Station 4) directly under the landing path, three fourths of the interviewees complained about having to make frequent home repairs, only to have the damage recur. Although vibration measurements were beyond the scope of this study, vibration caused by jet flyovers was very noticeable within the homes during several interviews in Tinicum.

A small but important number of residents believed that aircraft had directly affected the market value of their property. They made references to other people who had encountered difficulty in obtaining a fair market value for their homes when they wanted to leave Tinicum. Other respondents spoke from their own experience. They felt land-locked. This finding contradicts many studies on land-economics show-

TABLE 6.15.   Type of Property Effect

Category	Tinicum Township		West Philadelphia		Total	
	Number	%	Number	%	Number	%
Property devalued	14	17.95	1	1.69	15	10.95
Structural damage	32	41.03	12	20.34	44	31.12
Broken windows	13	16.67	1	1.69	14	10.22
Vibration	12	15.38	32	54.24	44	31.12
Other	7	8.97	13	22.04	20	14.59
TOTAL	78	100.00	59	100.00	137	100.00

Residential Effect of Noise and Perceived Noise Intensity

TABLE 6.16.

Answer	Perceived noise in neighborhood					
	Very quiet	Fairly quiet	Slightly noisy	Noisy	Very noisy	Total
No	56	145	88	38	18	345
Possibly	8	14	11	11	4	48
Yes	5	23	12	22	25	87

Chi Square 71.45491, Level of significance P<.001

ing that land values increase near airports. Land values do rise, but *usually when residential land is converted to commercial and/or industrial purposes.* This fact is little consolation for those wanting to maintain a single-family residential character in their neighborhood.

Considerable expenditures are necessary to maintain a positive residential environment in the vicinity of a commercial airport. Because of necessary insulation and navigational easement costs, experience shows that "property exposed to jet noise" (above 100 PNdB, or 87 dBA) "is worth 10–20 percent less than it would be if not exposed to jet noise."[3]

A significant chi square was found (P<.001) between the perceived effect of noise on property and the perception of neighborhood noise (see Table 6.16). Those believing that their property had suffered some type of damage due to noise understandably felt their neighborhood to be noisy, and consequently a less than desirable place to live.

Do people believe that the noise where they are now living is harmful to their health and well-being? A significant number, better than one in four (see Table 6.17) replied affirmatively, with another 3 percent considering noise to be possibly harmful. In Tinicum Township, these opinions were held by a majority. A lesser, but still notable percentage

HEALTH AND WELL-BEING

Noise Perceived as Harmful to Health and Well-Being

TABLE 6.17.

Category	Tinicum Township		West Philadelphia		Total	
	Number	%	Number	%	Number	%
No answer	1	1.00	12	3.00	13	2.60
Yes	58	58.00	71	17.75	129	25.80
No	39	39.00	299	74.75	338	67.60
Possibly	1	1.00	14	3.50	15	3.00
Don't know	1	1.00	4	1.00	5	1.00
TOTAL	100	100.00	400	100.00	500	100.00

(18%) of those in Philadelphia also thought community noise was a health hazard.

The relationship between the perceived intensity of noise and its believed harmful effect upon health was statistically significant in Tinicum (P<.01) as well as in West Philadelphia (P<.05). The noisier the neighborhood was rated, the greater likelihood that health was believed to be affected.

Two out of every five people interviewed stated that they felt differently after being exposed to noise. The typical effect was a change in behavior, rather than an organic or a physical change. Further questions led to identifying and categorizing these health-related effects, which are shown in Table 6.18. Nearly three fourths of the time, noise created a feeling of anger, irritability, or nervousness. Many people remarked that these reactions were seasonable, occurring more in warmer weather when noise from the out-of-doors was more noticeable. Only occasionally did anyone think noise had affected him physically, such as by a loss in hearing.

A noteworthy finding was that *the ability to hear is not a prerequisite of being disturbed by noise.* One totally deaf man conveyed through sign language the fact that he was unable to sleep because of vibrations caused by aircraft. These vibrations shook the entire house, including his bed, and consequently he was frequently awakened at night, even though he could not hear any noise.

CONCERN FOR NOISE ABATEMENT

In reply to our survey questions, respondents expressed concern for abating noise (see Table 6.19) and suggested a variety of methods (see Table 6.20). Many had already sought relief individually. Over one fourth of the respondents were using at least one method to reduce the intrusion of noise into their households. The method used most frequently was simply to keep windows and doors closed all year round.

TABLE 6.18.   Health Response to Noise

Health response	Tinicum Township		West Philadelphia		Total	
	Number	%	Number	%	Number	%
Angered–irritable	26	32.10	63	36.00	89	34.77
Nervous	31	38.27	63	36.00	94	36.71
Somatic disorder	9	11.11	19	10.86	28	10.94
Fatigue	2	2.47	8	4.57	10	3.91
Organic disability	6	7.41	1	.57	7	2.73
Other	7	8.64	21	12.00	28	10.94
TOTAL	81	100.00	175	100.00	256	100.00

Residents Using Noise Abatement Methods   TABLE 6.19.

Response	Tinicum Township		West Philadelphia		Total	
	Number	%	Number	%	Number	%
No answer	8	8.00	11	2.75	19	3.80
Yes	32	32.00	107	26.75	139	17.80
No	60	60.00	282	70.50	342	68.40
TOTAL	100	100.00	400	100.00	500	100.00

Next was the technique of masking exterior noise by using a radio, television, phonograph, air conditioner, or some combination of these. Still others had gone to the expense of adding sound insulation to their homes. Because people were concerned about getting adequate sleep, two methods were specifically suggested for this purpose: ear protection ranging from cotton to fitted earplugs; and medicine to induce sleep. A small number had adopted the practice of taking sleeping pills habitually. While most respondents concentrated on treating noise solely as it affected their homes, some resorted to social action by attacking the noise source directly. In Tinicum Township, a number of citizens had initiated a lawsuit against the City of Philadelphia, operators of the airport. People surrounding this facility had formed a noise abatement organization, the Tinicum Township Noise Abatement and Air Pollution Society. It is still active, although citizen interest varies.

In interpreting the acoustical and social data, it seems that individual recognition grows into community recognition as noise

Noise Abatement Methods   TABLE 6.20.

Abatement Method	Tinicum Township		West Philadelphia		Total	
	Number	%	Number	%	Number	%
Closing doors & windows	13	33.34	52	37.68	65	36.72
Other	3	7.69	22	15.94	25	14.12
Change Activity	1	2.56	17	12.32	18	10.17
Mask: Radio & TV	3	7.69	9	6.52	12	6.78
Mask: Air conditioning	3	7.69	14	10.15	17	9.60
Register complaint	2	5.13	10	7.25	12	6.78
Insulation	7	17.95	4	2.90	11	6.22
Ear protection	5	12.82	6	4.35	11	6.22
Medicine	2	5.13	4	2.90	6	3.39
TOTAL	39	100.00	138	100.00	177	100.00

increases in intensity and length. Those bothered by noise in West Philadelphia have devised mainly individual techniques to combat noise, while most Tinicum residents, who are closer to the airport runway, feel that larger-scale solutions are necessary.

A sizable number of the sampled population indicated that noise bothering them could be abated (see Table 6.21). The big question was who was responsible. In West Philadelphia those bothered by noise have devised some technique of their own to reduce the problem. In Tinicum this didn't seem to be happening. Less than half of the Tinicum respondents believed that any of their methods were solving the noise problem. Nearly 60 percent agreed that something could be done to reduce the noise bothering them, but just 32 percent were doing anything about it directly themselves. To the Tinicum people, solving the community noise problem, obviously due to jet aircraft, required large-scale solutions; the individual homeowner could do little to improve the situation. Furthermore, most homeowners there considered the problem to be the responsibility of the airport operators (City of Philadelphia) and their Township government.

A variety of solutions were suggested to reduce the noise on a community basis (see Table 6.22). The two most common suggestions from Tinicum residents were to relocate the population now residing in areas most affected by aircraft noise, and to remove this noise source by changing the flight patterns.

A lesser number stated that present aircraft technology could be improved upon. Many others thought that if this were possible (e.g., engine noise suppression), it would already have been tried. Home insulation was not mentioned, often, probably because the respondents believed they themselves would be financially responsible. It was also dismissed as an incomplete method of noise control, since the homeowners still could not enjoy their yards.

TABLE 6.21.   Response: Can Noise Be Abated?

	Tinicum Township		West Philadelphia		Total	
Response	Number	%	Number	%	Number	%
No answer	6	6.00	18	4.50	24	4.80
Yes	59	59.00	117	29.25	176	35.20
No	23	23.00	254	63.50	277	55.40
Don't know	12	12.00	11	2.75	23	4.60
TOTAL	100	100.00	400	100.00	500	100.00

Suggested Noise Abatement Methods   TABLE 6.22.

Methods	Tinicum Township Number	West Philadelphia Number	Total Number
Relocate people	19	4	23
Remove source	18	40	58
New regulation	11	13	24
Improve technology	7	9	16
Limit use	2	6	8
Sound Proof	2	5	7
Educate	2	4	6
Taxation: Offender	1	0	1
Taxation: Residents	0	1	1
Enforce existing regulations	0	17	17
Travel	0	6	6
Develop community activities	0	9	10
Parental control	0	110	110
Other	8	21	29

A significant number were disturbed to the point that they thought removing the source of noise was the only answer. Usually the West Philadelphia residents were referring to the noise of teenage activities in the neighborhood. However, a small number made reference to ground transportation (bus, trolley, and subway), feeling that these should be restricted in their operation. Some people living near playgrounds wanted them closed because they were used as "hangouts" by teenage gangs. Many of those suggesting the "enforcement of existing regulations" made reference to the city's existing noise ordinance, a regulation more or less on paper. Still other people, not knowing of this ordinance, recommended that one be adopted. For temporary relief, traveling was suggested as a weekend escape from the noise of the city. Many felt that the noise caused by children could be controlled, in part anyway, by parents exercising greater discipline. Others thought that the community should develop and sponsor more activities for children.

It is interesting also, to examine the respondents who, although bothered by noise, felt that noise couldn't be abated (see Table 6.23). A commonly expressed opinion was that noise is a "by-product of progress" and therefore we must live with it since "progress cannot be tampered with." Another popular opinion was that the scale of the noise problem was so large and so complex that effective solutions were highly unlikely.

Only one person in ten remarked that he or she would definitely

TABLE 6.23.   Why Nothing can be Done to Reduce the Noise Bothering the
               Individual

	Tinicum Township		West Philadelphia		Total	
Response	Number	%	Number	%	Number	%
No answer	4	26.67	71	43.55	75	42.16
By-product of progress	1	6.67	41	25.15	42	23.60
Abatement not possible	3	20.00	7	4.30	10	5.62
Scale of the problem	4	26.67	39	23.93	43	24.16
Know no method	3	20.00	5	3.07	8	4.46
TOTAL	15	100.00	163	100.00	178	100.00

bear part of the cost of noise control (see Table 6.24). Those considering
their neighborhood a noisy environment were least inclined to bear
even part of the cost to reduce unwanted sound. In their words, *they
had not caused the problem, therefore they were not responsible for controlling it.* This was the prevailing opinion in Tinicum, where, in contrast with most airport communities, the population remains stable.
The people are long-term residents, averaging nearly 25 years' residence
in that community. Many have observed the airport's growth from a
small grass landing strip serving two-engine propeller aircraft to a
large modern facility supporting heavy jet traffic. Other reasons were
given besides not being responsible for the noise (see Table 6.25). Some
felt they could not afford any additional expenses of this type, or that
they were already paying what they considered their fair share. The
largest percentage of the interviewees were not bothered enough to
contribute personally to the control of community noise.

The most acceptable method of financing community noise control,
according to the few definitely or possibly willing to pay, was by
voluntary contributions (see Table 6.26). Direct taxation was only half

TABLE 6.24.   Willingness to Bear Part of the Cost to Reduce Noise in the Area

	Tinicum Township		West Philadelphia		Total	
Response	Number	%	Number	%	Number	%
Yes	10	10.00	39	9.75	49	9.80
No	40	40.00	77	19.25	117	23.40
Possibly	44	44.00	267	66.75	311	62.20
No Answer	6	6.00	17	44.25	23	44.60
TOTAL	100	100.00	400	100.00	500	100.00

Reasons for Unwillingness to Bear Part of the Cost to Reduce Noise   TABLE 6.25.

	Tinicum Township		West Philadelphia		Total	
	Number	%	Number	%	Number	%
Can't afford to	11	30.56	40	26.14	51	26.98
Not bothered enough	8	22.22	72	47.06	80	42.33
Not responsible	12	33.33	13	8.50	25	13.23
Nothing can be done	4	11.11	17	11.11	21	11.11
Already paying	1	2.78	11	7.19	12	6.35
TOTAL	36	100.00	153	100.00	189	100.00

as popular. Those responses classified as "other" represented a variety of ideas: from fund-raising projects such as raffles, auctions, or benefits to levying a surcharge on passenger tickets (i.e., airline, bus, train). The average amount people were willing to pay annually was between five and ten dollars.

COMPENSATION

Far from being inclined to pay for noise abatement, a sizable number, primarily in Tinicum (36%), thought they were entitled to financial compensation (see Table 6.27). This opinion was expressed in response to the question, "If your local noise problem cannot be solved, do you believe you should receive some financial compensation?" As would be expected, a higher proportion perceiving their neighborhood as noisy desired compensation. There were, however, departures from this general tendency. Monitoring Station 4, having the highest flyover sound levels, did not have the largest number of respondents wanting compensation. Nor is the desire to be compensated attributable to only one type of noise, although aircraft noise leads the list. In Philadelphia, where just over 9 percent wanted compensation, various noises (vehicular, trolley, train, bus, subway and general) were responsible.

Methods of Paying for Noise Abatement   TABLE 6.26.

	Tinicum Township		West Philadelphia		Total	
Response	Number	%	Number	%	Number	%
Taxes	10	28.57	26	30.95	36	30.25
Voluntary contribution	21	60.00	45	53.57	66	55.46
Other	4	11.43	13	15.48	17	14.29
TOTAL	35	100.00	84	100.00	119	100.00

TABLE 6.27.   Perceived Entitlement to Compensation

Response	Tinicum Township		West Philadelphia		Total	
	Number	%	Number	%	Number	%
No answer	18	18.00	79	19.75	97	19.40
Yes	36	36.00	37	9.25	73	14.60
No	42	42.00	278	69.50	320	64.00
Don't know	4	4.00	6	1.50	10	2.00
TOTAL	100	100.00	400	100.00	500	100.00

The amount of compensation desired varied from less than $100 to over $5,000 (see Table 6.28). A sizable number of responses were classified as "don't know" for one of two reasons. Some of the Tinicum homeowners were seeking relief through the courts. The compensation these people were demanding was based on the loss in fair market value of their property. At the time of the survey, real estate appraisals had not yet been completed to determine the amount sought by each claimant.* Other people wanted to have building repairs made or sound insulation added, but they did not know what the costs might be. In analysis by chi square, *a consistent relationship did not appear (P>.05) between the variables of perceived noise and the amount of compensation.*

Social Demographic Analysis

A series of social demographic factors was included in the questionnaire in order to determine if these factors affected the way people

TABLE 6.28.   Desired Dollar Value of Compensation

Response	Tinicum Township		West Philadelphia		Total	
	Number	%	Number	%	Number	%
Under $100	0	0.00	9	25.71	9	14.28
$100–$499	0	0.00	5	14.29	5	7.94
$500–$999	1	3.57	2	5.71	3	4.76
$1,000–$1,999	2	7.15	8	22.86	10	15.87
$2,000–$2,999	1	3.57	0	0.00	1	1.59
$3,000–$3,999	1	3.57	0	0.00	1	1.59
$4,000–$4,999	1	3.57	3	8.57	4	6.35
Over $5,000	1	3.57	0	0.00	1	1.59
Don't know	21	75.00	8	22.86	29	46.03
TOTAL	28	100.00	35	100.00	63	100.00

*Plaintiffs have sought compensation amounting to 15% of the property's market value.

perceived the noise problem in their community. If there was a high degree of correlation between demographic characteristics and noise perception, then this analysis could be helpful in formulating a framework for understanding and possibly predicting human response patterns. The base variable required the respondent to rate his neighborhood according to one of five categories. The question asked was "Do you consider your neighborhood to be very quiet, fairly quiet, slightly noisy, noisy, or very noisy?"

HOUSING

Analysis by chi square techniques indicates that the respondent's type of housing is related to his perception of noise as a problem (P<.01). It appears that the responses of the rental population differs appreciably from those of the owner population (see Table 6.29). In our study, those owning their homes expressed greater concern about noise than did apartment dwellers. A partial explanation may be that the apartment resident is more mobile and less financially committed to a fixed location than a homeowner. He is able to move more easily if living conditions become a serious problem. Furthermore, among other considerations, the renter does not have to be concerned about noise affecting the value of his property.

AIR CONDITIONING FOR NOISE CONTROL

Regarding techniques for controlling the penetration of noise, it is apparent that residents who have air conditioning (central or window units) perceive community noises as less of a problem than those who have no air conditioners. There is, however, a notable exception. Around the airport, where flyover noise was found to be very intense, the use of air conditioning brought little or no relief to those answering the questionnaire. A comparison of the opinions of those having storm windows (or window fans) with the opinions of those who have no such devices in Tinicum showed that these did little to make the noise subjectively more tolerable.

Type of Dwelling Unit and Perceived Noise in Neighborhood

TABLE 6.29.

Type of dwelling unit	Perceived noise in neighborhood					
	Very quiet	Fairly quiet	Slightly noisy	Noisy	Very noisy	Total
Apartment	11	43	20	12	3	89
Flat	0	6	3	1	1	11
Row	34	95	71	36	19	255
Semi-detached	14	18	5	5	9	51
Single-family	14	26	15	17	12	84
TOTAL	73	188	114	71	44	490

Chi square 33.57432   Level of significance (P<.01)

AGE OF RESIDENTS

If today's popular music is any indication, modern youth are more tolerant of noise than are adults. When teenagers were present while our questionnaire was being answered, their attitudes toward noise usually varied from those of their parents. The head of the household, or his spouse, predictably was more annoyed. Unfortunately the younger generation was not included in the sample design; thus conclusive facts concerning their views are not available.

Considering the adult group interviewed, it seems that age did not affect the way noise was perceived (P>.10). The differences among age groups seemed slight (see Table 6.30). The chance that age mattered seems even smaller in the Tinicum area than in West Philadelphia. If noise is intense enough, all adults, irrespective of age, are generally affected.

Many Tinicum mothers made side comments that when their children were young babies they could not adjust to aircraft noise, even after months of exposure. These babies usually awakened during their sleeping period upset and often crying. Only when they became toddlers was the sleeping problem overcome.

MOBILITY OF
RESIDENTS

Before undertaking our survey, we had believed that community noise exposure and the degree of human response are related to individual mobility. However, significant relationships were not found. The length of time our respondents lived at one address had no effect on their noise perception. People living in their homes for a short time responded much as did long-term residents (see Table 6.31), whether in Tinicum or in West Philadelphia. This result leads us to believe that people do not adapt in time to community noise.

The length of time spent daily in the neighborhood was not related to noise perception. Persons spending eight or more hours daily outside the area did not differ significantly in their reaction to noise from

TABLE 6.30.   Age and Perceived Noise in Neighborhood

| Age of respondent | Perceived noise in neighborhood | | | | | |
	Very quiet	Fairly quiet	Slightly noisy	Noisy	Very noisy	Total
Under 20	3	2	2	2	–	9
20–30	14	32	22	9	11	88
31–40	6	39	26	17	10	98
41–50	11	32	21	9	4	77
51–60	14	33	20	19	15	101
61–70	13	29	13	9	6	70
70+	10	16	9	6	–	41

Chi square 29.47012   Level of significance P>.10 (not significant)

Number of Years at Present Address and Perceived Noise in
Neighborhood

TABLE 6.31.

Length at present address	Very quiet	Fairly quiet	Slightly noisy	Noisy	Very noisy	Total
Under 1 yr.	9	13	6	6	2	36
1–3 yrs.	1	10	6	4	1	22
4–6 yrs.	6	14	8	5	1	34
7–10 yrs.	2	16	12	11	9	50
11–15 yrs.	6	20	12	7	9	54
16–20 yrs.	11	27	20	10	5	73
21–25 yrs.	5	6	12	7	8	48
26–30 yrs.	19	44	21	14	7	105
Over 30 yrs.	13	25	17	7	4	66

Chi square 32.51085   Level of significance P>.10 (not significant)

those who were home all day. The apparent conclusion is that the
threshold of annoyance varies according to time of day. The most
critical time is during the evening and night hours. It is evident that
people are more annoyed by noise during their hours of relaxation than
during normal working hours.

LIFE STYLE

Where an individual had lived during his childhood had an
effect upon his judgment of his adult environment. A question in
our survey asked, "During most of your years, up to the age of 18, where
was your home?" The replies, classified according to urban, suburban,
and rural environments, were compared with perceptions of current
noise levels. People who had previously lived in either rural or suburban communities considered their current location in Philadelphia to
be noisier. The same held true in Tinicum. In Tinicum, even people
with a city background felt in nearly every instance that their present
adult environment was noisier. West Philadelphia respondents felt
that it was at least as noisy or noisier. Generally, however, people
raised in city environments were less critical of noise than others.

OTHER VARIABLES

Additional demographic variables were evaluated, including education, race, sex, and income of the respondent. There appeared to be
no statistically significant correlation between these variables and the
perception of neighborhood noise.

VII

Quieting the Crisis:
Some Solutions

Field study results strongly indicate that the health of the urban population, in its broadest sense, is being compromised by the presence of noise. Noise attacks on health are of various degrees of severity, ranging from hazards to nuisances. Most health problems related to community noise fall into the latter category (i.e., mental stress, task communication and sleep interference, and general annoyance). Community recognition of the problem appears to be fairly widespread. However, human responses are not always proportional to noise intensity. Besides, certain community noise sources with similar sound pressure levels evoke different degrees of annoyance.

Disregarding for the moment the many refinements that should be made in the methodology and analysis of the study of community noise, a basic health rating system can now be constructed as a result of our field study, which would allow us broadly to assess the impact of noise upon the quality of environment (see Table 7.1). Such an analytical tool begins to provide a comprehensive health picture of the environment at any scale, depending upon its application (e.g., household, neighborhood, community, state). With the results of this rating system, noise management problems can be more easily identified, to facilitate the decision-making process.

The environment can be evaluated with the aid of nine health indexes symptomatic of noise pollution: vascular constriction, hearing loss, mental stress, sleep interference, task interference, communication interference, property damage, friendship formation and/or maintenance, and annoyance. These indexes cover the potential range of

A System for Rating Environmental Health

TABLE 7.1.
Environmental Health Rating System: Noise

HEALTH INDICES	MAGNITUDE OF EFFECT															
	SURVIVAL				INJURY				EFFICIENCY				COMFORT & ENJOYMENT			
	Value	*Time*	*Pop*	*Total*	*Value*	*Time*	*Pop*	*Total*	*Value*	*Time*	*Pop*	*Total*	*Value*	*Time*	*Pop*	*Total*
Vascular Constriction																
Reversible									2				1			
Irreversible	4				3				2				1			
Hearing Loss																
Temporary	4				3				2				1			
Permanent	4				3				2				1			
Mental Stress	4				3				2				1			
Sleep Interference									2				1			
Task Interference									2				1			
Communication Interference																
Face-to-Face	4				3				2				1			
Telephonic	4				3				2				1			
Property (Artifacts)																
Damage	4				3				2				1			
Destruction	4				3				2				1			
Friendship Formation and/or Maintenance									2				1			
Annoyance									2				1			

effects of noise, from a threat to human survival and physical injury to reduced efficiency and decreased personal comfort and enjoyment.

The more hazardous a noise effect is to human health, the more it will be weighted in the rating system, on a scale from one to four. This weighting is corrective: a noise that challenges human survival is weighted four times more than one that causes only subjective discomfort or a loss of enjoyment. As an example, communication interference. If the inability to hear voice communication because of intruding noise endangers human life, this noise is considered four times as important as the noise which is annoying to social conversation. The value of the effect will be termed V.

The length of time for which health is affected, and the portion of the population affected, provide the remaining quantitative indexes. Any figure less than 1.0 signifies that only a portion (or percentage) of time or population is being affected. An overall total number is obtained for the health index at each level of effect by multiplying the corrective value with time, and again with population: total $= V \times T \times P$.

This formula can be applied to any of the nine indexes found in Table 7.1. Consider sleep interference. First of all, sleep interference constituted a health problem to the extent of affecting either efficiency or comfort and enjoyment. Efficiency becomes a health problem during the sleeping process at approximately 55–65 dBA (inside the home), because a statistically significant portion of the population is awakened, and there is a loss in REM sleep at this sound level and above.[1] The recuperative value of the sleep that is necessary for body maintenance is reduced, since disruption of sleep results in temporary physiological and psychological changes of state. These changes contribute to a reduction in human efficiency. Between 35 and 45 dBA is the threshold for physiological change where brain wave patterns (EEG) are altered, though the subject does not necessarily awaken.[2] It is at this level that noise begins to affect human comfort physiologically.

With the data collected in our field study, the significance of community noise and sleep loss can now be evaluated. Consider Monitoring Station 4 in Tinicum Township, a high sleep risk population because of the adjacent airport. During the nighttime operating hours (11:00 P.M.–7:00 A.M.), there were at this location approximately 38 flyovers (usually landings), during every hour of the night. Nearly all were by jet craft, so that the sound level outside the dwelling unit averaged 90–105 dBA (inside, the level was approximately 70–85 dBA). These intensities affect human efficiency, as answers to our social questionnaire have established.

Understandably, the percentage of the sample population affected was sizable. In fact, all respondents at Station 4 indicated that they were annoyed by these overflights during those hours. Frequency of disturbance was the major variable, with 65 percent feeling disturbed often, and the remaining 35 percent feeling disturbed sometimes in their sleep. To all intents and purposes, the entire Station 4 population was at risk. It is more difficult, however, to determine exactly what percentage of the eight-hour sleep period was disturbed by this intrusion. Each overflight lasted approximately thirty seconds. The minimum period of disturbance for 38 overflights was therefore nineteen minutes. In addition, there was the time required to return to the previous stage of sleep (regardless of whether or not a person awakens). Many people awakened completely; they often got up and read, ate a snack, went to the bathroom, or just lay awake momentarily before resuming their sleep. *On the average,* approximately forty minutes a night was required to "recover" from such flights, although this is a highly variable estimate. The total period of sleep interference at this station averaged nearly sixty minutes (i.e., noise onset 19 minutes, and recovery 40 minutes) or roughly one hour per night.

This would complete the information necessary to calculate sleep interference and efficiency. In our example, the population exposed was found to be 100 percent, the length of interference 12.5 percent of the sleeping period, based on eight hours of sleep per night. The total is .250, with $V = 2.0$, $T = .125$ and $P = 1.00$.

Similar estimates can be made for the remaining eight indexes at this station. When all indexes are evaluated, the result is a comprehensive environmental health profile of noise pollution as it affects a given population at a given location. The same evaluational technique can then, for comparative purposes, be applied to other sampling areas. Once this is done, households, neighborhoods, communities, and larger political-geographical units can be ranked according to their environmental health status. Solutions and priorities can thus be established for minimizing the health effect of noise upon the environment.

Community Noise Management: Components

The proper management of community noise requires a study of the three components of the problem: the source, the path, and the ultimate receiver of the noise (see Figure 7.1). Failure to consider any one of these components will reduce the effectiveness of any proposed solution.

Treating noise at the source is clearly the most desirable approach. However, apart from the fact that this solution is sometimes technologically and economically impractical, communities may not have the necessary legal jurisdiction. The control of navigable air space, for

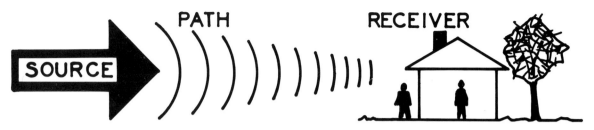

FIGURE 7.1. *Noise management components*

example, including the noise generated by aircraft in flight, is exclusively a federal government responsibility. Local and state governments do not have legal authority to set aircraft noise limits.

A second approach is to treat the path along which noise travels, rather than the source of the noise itself. Techniques include noise reduction by diversion, reflection, absorption, and dissipation (increasing the distance between the source and object). The last resort, and often the least desirable solution, is to treat the noise at the site of the receiver. In communities this means either protecting people individually or treating structures occupied by people. Every community has a variety of means to institute a noise management program involving source, path, and receiver treatment or combinations thereof.

RECEIVER/OBJECT

Once a noise reaches the individual, and a potentially health-hazardous condition exists, few options are open other than wearing individual hearing protection (i.e., earplugs or earmuffs). Such apparel certainly detracts from using, much less enjoying, the environment. Fortunately, before such an extreme solution becomes necessary, methods are usually available to control the use of the land and the buildings affected.

Building codes, because they cover construction requirements, are important devices for establishing acoustical controls. Special requirements should be established in intense noise areas (the vicinity of airports, etc.). Only one major municipality in the United States, New York City, has a building code which contains noise control requirements.[3] Even this code has two notable shortcomings: it applies only to multiple family dwellings, and "the provisions are not directed at protection against noise emanating from outside the building."[4] Construction materials and methods to minimize noise intrusion would seem to be available, but an agency is needed to insure that these materials and methods are applied. Building and housing codes are a partial answer, at least with new building construction. As for existing housing, acoustical materials and methods are available to modify these structures also; but there is little initiative or incentive to change current building practices.

The U.S. Department of Housing and Urban Development has issued a circular to guide the various HUD programs in noise abatement and control.[5] This circular is primarily oriented to proposed rather than existing structures. There are some interior noise standards for rehabilitated residential construction. Noise levels for sleeping quarters are "acceptable" if they do not exceed the following standards:

1. do not exceed 55 dB(A) for more than an accumulation of 60 minutes in any 24-hour period, and
2. do not exceed 45 dB(A) for more than 30 minutes during nighttime sleeping hours from 11 P.M. to 7 A.M., and
3. do not exceed 45 dB(A) for more than an accumulation of eight hours in any 24-hour day.[6]

In most areas of the field study at least one of these HUD standards would be exceeded, and near the airport all three noise standards would be exceeded. An additional requirement for proposed multifamily structures is a sound insulation standard. Insulation between dwelling units to be acceptable must have a Sound Transmission Class (STC) of at least 45.[7]

Zoning can be effective, because it can minimize the number of persons exposed to noise by regulating building height, bulk, and density. Front- and side-yard zoning requirements in residential districts, though not used for this purpose, are useful in limiting noise intrusion from outside a dwelling. Zoning controls are most effective in treating the path of noise (see the next section).

Interior site planning is another method of minimizing noise. "Windows are the weakest part of exterior walls in terms of sound control," so care should be exercised in choosing their placement, size, and number.[8] For a municipality to incorporate interior site planning requirements into a housing code might be legally questionable, but a listing of recommended site planning guidelines for noise control would be very useful. When land being developed is under the control of a housing authority, general acoustical requirements can be stipulated. This has been done very successfully in London by the Housing Authority, and without architectural compromise. In preparing an acceptable housing scheme, a British architect had to obviate the major roadway and aircraft noise problem near London's Heathrow airport. The design solution for Heston Grange, a rental housing development, proved esthetically and acoustically pleasing.[9]

Formal controls are not always necessary—public pressure can bring about change. In the San Francisco Bay area, certain high-rise rental housing was boycotted because the public complained that it was too noisy. People refused to rent these units, and so the developer hired an acoustical consulting firm. Significant improvements were incorporated to make the units more habitable. The same consulting firm has been retained by the developer on subsequent high-rise apartment projects, which thus gained wider public acceptance.[10]

Many important solutions require that the noise be intercepted between its source and its receiver. Most of these solutions involve some type of land management control. Comprehensive planning in conjunction with zoning can be a formidable deterrent to the spread of community noise in residential areas. Land uses that are incompatible because of noise should be separated. Distance, as mentioned earlier, is a primary means of reducing noise; normally when distance is doubled, noise is reduced by six decibels. If planning and zoning are initiated early enough and are not compromised, then noise generators such as airports can be effectively isolated from sensitive residential areas. There exist several municipal zoning ordinances setting noise performance standards to regulate industrial land uses.

PATH OF NOISE

Review of subdivision regulation is another possible planning technique. As part of the review, the developer could be required to prepare noise contours, or at least representative noise measurements, of the site being developed. Site design solutions would have to include noise control provisions if the prevailing or anticipated noise within the site would exceed a prescribed sound level. HUD has proposed a series of site standards for new residential construction. A site is considered unacceptable when at an appropriate height above the site boundary the noise level exceeds:

1. 80 dB(A) for over 60 minutes in any 24-hour period
2. 75 dB(A) for over 8 hours in any 24-hour period.[11]

At the present time these standards are not being implemented.

In ground transportation planning, roadway "improvements" or new roadways should require noise control solutions if they appreciably change the acoustical character of the adjacent area. Various barriers (e.g., walls, nonresidential buildings, elevation and depression, embankments) can alter the path of noise and thereby reduce the amount of undesired sound reaching the population.[12,13]

Noise nuisance regulations enacted by local and state governments attempt to control noise at its source. Such regulations place the burden on the user rather than on the community or public at large. Construc-

SOURCE OF NOISE

tion curfews, which limit the hours of the day when construction activity is permitted, are common. Vehicle noise emission limits are receiving more and more attention, particularly from state governments. Indirectly, these limits put pressure on automobile manufacturers and their suppliers to design and engineer a quieter product. (According to some engineers in industry, if the noise levels stipulated in the California Motor Vehicle Code are lowered further, certain automobiles currently being manufactured will all be in violation.) Aircraft noise associated with ground operations (for example, engine run-up, engine testing) is also becoming the subject of many community ordinances.

Since all governments are consumers, their buying habits play a major role in determining how severe the noise problem is at present and what it will be in the future. By enacting a policy of "buying quiet," a city can set an example for others to follow. Furthermore, the manufacturing industry can be re-educated as to the needs of the consumer, although the consumer must be prepared to pay for the amenity of quiet. Governments should incorporate noise specification requirements into purchase orders for those items which create a community disturbance (for example, street maintenance and repair equipment, public transportation conveyances, construction equipment). The more unified and comprehensive the effort, the lower the unit cost. (Efforts such as those in New York City are commendable, but they need to be extended throughout the country.)

The U.S. Army has instituted this policy on many items they procure. Most specifications concerning materiel that may affect the health of Army personnel are reviewed by the U.S. Army Environmental Hygiene Agency. This agency prepares noise control requirements for those specifications when noise could be a health hazard.

MANAGEMENT OF
AIRCRAFT NOISE

Data from our acoustical survey have identified certain noise sources within the residential areas sampled that need to be controlled. These sources represent the principal contributors to the current level of community noise. If a noise management program is directed primarily toward solving the problems of transportation noise, then the din of our cities will be on its way toward being controlled.

Both in terms of community opinion and comparable noise measurements, aircraft noise stands out as the most significant problem. It is very obvious that a sizable portion of the population surveyed has become intolerant of aircraft noise. However, there are several possible solutions that need to be explored in order to bring relief to the affected population, especially those living in Tinicum Township. The new parallel runway under construction is only a partial answer. It *may reduce, but will not eliminate the intense aircraft noise to which the population is exposed* (see Chapter IV). Furthermore, until that runway becomes operational, the entire Tinicum population, and to a lesser extent Philadelphia residents, will continue to be affected.

Not all the noise management solutions presented fall within the local purview, but the majority of them can be applied by the City of Philadelphia, operator of Philadelphia International Airport.

An increase of the current glide slope angle above the 3° now used by aircraft approaching for a landing would lessen the area and intensity of community noise exposure. For example, if the landing profile used were a 6° rather than a 3° glide slope (see Figure 5.14), there could be a reduction of approximately 7–8 PNdB. A reduction occurs since the aircraft is at a higher altitude and the power setting is somewhat lower.[14] Because of aircraft and passenger safety, authorities believe 6° to be the maximum possible glide slope. Although this landing profile offers tangible relief from the noise, it is not being used.

SOURCE

Landing Profile

A takeoff procedure has been developed by the Federal Aviation Agency for minimizing aircraft noise in populated areas. Referred to as a standardized noise abatement takeoff profile, it requires jet aircraft operators to follow these procedures after liftoff:[15]

Takeoff Profile

(1) accelerate to V2 plus 10 to 20 knots with takeoff flap and takeoff thrust;
(2) at no sooner than 400 feet, initiate flap retraction schedule and accelerate while maintaining positive climb gradient, achieve minimum maneuvering speed and clean configuration by 1500 feet;
(3) stabilize with minimum maneuvering speed and target EPR;
(4) at 4,000 feet, resume en route climb schedule.

Such a procedure is being used at Washington National Airport in cooperation with the FAA, Air Line Pilots Association, and the Air Transport Association. Noise levels on the ground are thereby reduced by 8.8 PNdB below those under existing airport procedures elsewhere.[16] In Philadelphia, this procedure is not being used.

Because of community response, several major airports have restricted or eliminated nighttime aircraft traffic. Curfews are in effect at London's Heathrow Airport and at least nine airports in the United States.[17] After 11:00 P.M. no scheduled commercial flights are permitted at Washington National, or at O'Hare in Chicago. In England, the Ministry of Aviation strictly limits flight operations at the Heathrow Airport during late evening and night hours.[18]

Curfews

The legality of airport operators restricting commercial jet flights has not been fully tested in the courts. Lower and intermediate courts have allowed small general aviation airports to establish curfews.[19,20] The situation may be different with major airport hubs. Such regulation could be in conflict with the flow of interstate commerce. The FAA has the authority, but has not chosen to apply a curfew policy in Philadelphia. For obvious economic reasons the airport administrator would be happy to get any additional flights during off-peak hours, regardless of the time of day.

Establishing
Permissible Noise
Levels

Some airports have imposed noise restrictions on aircraft during takeoff. At JFK in New York there is a limit of 112 PNdB, while in London there is a daytime limit of 110 PNdB, decreasing to 102 PNdB during the evening hours.[21] Two general aviation airports (i.e., Santa Monica, California[22] and Morristown, New Jersey[23]) also have established permissible noise levels for takeoff operations. Again, the legality of such action has not been fully reviewed in the courts. Airports have avoided establishing limits for landing jet aircraft, probably because of safety considerations.

Restricting the noise of commercial jets on the ground is an entirely different matter. Ground run-up noise and engine testing limitations are found in several community ordinances. Legally this has been judged to be permissible. The City of Inglewood, which adjoins Los Angeles International Airport, specifies in their ordinance:

> It shall be unlawful for any person to operate, run up or test or cause to be operated, run up or test an aircraft jet engine which creates a noise level of 50 dBA or more at any place within an inhabited residential zone of the City of Inglewood between the hours of 10:00 P.M. and 7:00 A.M.[24]

No noise ceiling has been established in Philadelphia for either ground or air operations.

Enforce Existing
Flight Procedures

The Philadelphia airport management, in response to individual complaints, has insisted that landing aircraft follow the watercourse (Delaware River) exclusively in their flight approach. Although this may well be the intention, in actuality the aircraft fly over the populated area, sometimes extending as far north as 2nd Street (1,500 feet, or nearly one third of a mile inland). Such flight irregularities have frequently been observed during our survey in Tinicum. (The citizens also indicated this problem in the questionnaire.) These variations occur especially at night.

Encroachment over the land area heightens the noise problem. An error in lateral distance by an approaching aircraft affects both noise propagation and community annoyance. For example, an approach

1000 feet north of the prescribed flight landing path can raise the noise level over these homes by as much as *20 dBA.*

If the flight procedure is to follow the Delaware River on approach, then this procedure should be strictly monitored. But as there is pilot variability, any "error" should be made to the south, so that the aircraft are further over water rather than over the community. Appropriate lighting should be installed (as has been done in Washington, D.C.) to assist pilots in staying over water during their approaches from the west.

Airlines as tenants must pay a set fee for using an airport, including Philadelphia International. The noise problem could be improved if this fee were adjusted upward for certain periods of the day, particularly involving late evening and night flights.

Setting Landing Fee Schedule

Such a "tenant noise surcharge" is aimed at economically discouraging certain aircraft operations at the times most sensitive to community noise. It would tend to stabilize and insure periods of quiet that are important to the local residents for environmental health reasons.

Not only time would be a criterion for applying the surcharge, but so would type of aircraft. In other words, the greater the noise propagation, the higher the fee; jet aircraft would thus be most affected. The much less annoying propeller and turboprop-assisted aircraft, on the other hand, would not be burdened by such a surcharge—a reward for operating quietly.

Many businesses, when letting a contract, write into it certain specifications that must be met to satisfy the contract. These specifications or requirements can often be considered performance standards. Today, few noise standards are associated with industrial products; however, some of the larger industries and municipal governments require that the products they order must not exceed certain noise levels.

"Buy Quiet"

No such practice prevails in the aircraft industry. Consequently, producing "quiet" gives manufacturers no competitive edge. In 1968 when a manufacturer was selected to supply jet engines for the future airbuses, Britain's Rolls-Royce lost out to America's General Electric,[25] even though their engine, equipped with three rather than two turbine shafts, was judged to be quieter.[26] (Lockheed was the only one of the three airbus competitors to buy the British engine.) In order to insure less din in the future, the airlines, airport landlords, and others in authority, must assert their influence—and one way to do so is to specify and buy quiet when it is available. Ultimately this may require passing these costs on to the airport user.

Aircraft Noise
Easements

One HUD official has offered a scheme for compensating those affected by noisy overflights. The airport, instead of having a permanent right to make noise, would lease this right from landowners for a stipulated length of time. In the lease arrangement a maximum permissible noise level would be agreed to. If applicable, claims could be filed by the landowners at the time the leases needed renewing; then the settlement could be based on the number of overflights violating the noise level. An extensive series of noise monitoring stations with proper acoustical equipment would have to be established to insure compliance.

In Tinicum, which has only a small number of residents, a proposal like this could be manageable; however, in order for it to be most efficient, the local government should be the leaseholder. Revenue from the leasing of such an easement could be returned to the affected homeowners in the form of lower mortgage rates, a sound insulation program, lower taxes, and so forth.

Engine Suppression

Suppression of jet engine noise through acoustical engineering is the most direct route to quieter engines, but at present technological limitations seem to impede significant advances. Former Secretary of Transportation Alan S. Boyd, commenting on the potential of attenuating engine noise, said, "I do not believe there will ever be such a thing as a quiet airplane."[27] According to several authorities, advances are likely to be minimal in future aircraft, because of increased engine size and larger payloads.

However, improvements should be made that would change more quickly our existing stock of jet aircraft. Economic incentives should be applied to reduce the time lag. The Aerospace Industries Association reports a "projected" reduction of 3–11 PNdB (depending on aircraft size, etc.) on planes entering service between 1969 and 1972. There is little if any hope for those planes currently in service, without an engine retrofitting program. A retrofit program would be an expensive undertaking, exceeding $1 billion. Modifications for the present DC-8 would cost nearly $700,000 and about $11 million for the Boeing 707.[28] Even with these engineering modifications, takeoff noise would remain nearly the same. The only significant improvement would be landing noise.

PATH OF NOISE

Solutions involving the path rather than the source of noise shift noise management away from the aircraft itself. In this approach jet aircraft becomes the given factor, while land becomes the variable, since any noise reduction now requires altering the land. The aircraft interests prefer that the noise problem be dealt with from this angle, since the responsibility is then removed from them and placed upon the community. Airline profits are thus protected against diversion into additional research-and-development programs.

The city planning profession possesses several tools that can help minimize the airport noise problem, but they cannot wholly eliminate it.

At the present time no government agency knows, much less cares about, the number of land parcels, households and/or people currently being affected. In the area of our case study, community noise is not officially recognized as a problem needing attention by either the City of Philadelphia or Delaware County.

Land Use Inventory

A very important first step is to prepare a land use inventory of that population affected, using some noise criteria. HUD and DOT are jointly supporting grants to several metropolitan areas for the purpose of defining, in part, the present and future airport noise problem.[29] Land use and related data are being compiled; as their noise criterion they use the Noise Exposure Forecast. This NEF is essentially a method designed to predict, on the basis of certain acoustical and operational aircraft information, the degree of community annoyance.[30] The FAA has adopted this method for evaluating all airport community noise problems. Beginning at a certain point, the higher the NEF, the greater the annoyance and the less compatible the land use. For example, when the NEF falls below 30, land used for residential purposes is compatible. Between 30 and 40 NEF only, multiple family housing is compatible, and when the NEF exceeds 40, no residential land use (single or multiple family) is compatible. Though the recommendation is subject to question, industrial and outdoor recreational land uses are still permitted above a 40 NEF (see Table 7.2).[31]

Applying these NEF criteria to the data compiled by our field survey, aircraft using either the *present or the proposed runway configuration will render certain residential areas incompatible* (see Table 7.3). The present runway takeoffs at night (10:00 P.M. to 7:00 A.M.) create an NEF of 40 in approximately 80 percent of the residential area; during the day the percentage affected declines slightly, to 60 percent. *All single family units are incompatible,* since present takeoff noise generated by jet aircraft exceeds the 30 NEF. This percentage will remain the same even when the new runway opens in late 1971 or early 1972. With landing aircraft the noise problem is less severe for both the present and the future runway. Smaller residential (single-family) areas are and will continue to be incompatible, however.

Although the Noise Exposure Forecast may reliably predict annoyance due to aircraft noise, *such annoyance is not a consistent index of environmental health-related problems.* The Department of Transportation Office of Noise Abatement should not rely solely on the NEF method for assessing airport community noise problems. Health protection

TABLE 7.2.   Land Use Compatibility Chart for Aircraft Noise

Noise exposure forecast areas	Residential	Commercial	Hotel, motel	Offices, public buildings	Schools, hospitals, churches	Theaters, auditoriums	Outdoor amphitheaters, theaters	Outdoor recreational (nonspectator)	Industrial
A <30	Yes	Yes	Yes	Yes	Note (C)	Notes (A) and (C)	Note (A)	Yes	Yes
B 30 – 40	Note (B)	Yes	Note (C)	Note (C)	No	No	No	Yes	Yes
C >40	No	Note (C)	No	No	No	No	No	Yes	Note (C)

Note (A) A detailed noise analysis should be undertaken by qualified personnel for all indoor or outdoor music auditoriums and all outdoor theaters.

(B) Case history experience indicates that individuals in private residences may complain, perhaps vigorously. Concerted group action is possible. New single dwelling construction should generally be avoided. For apartment construction, Note (C) applies.

(C) An analysis of building noise reduction requirements should be made and needed noise control features should be included in the building design.

Source: Bolt, Beranek, and Newman, Inc., "Procedures for Developing Noise Exposure Forecast Areas for Aircraft Flight Operations," August 1967.
Prepared under contract with the FAA.

requires more than merely the removal of possible annoyance, since a portion of the population will not respond to community noise intrusion regardless of intensity. The use of the environmental health rating system suggested earlier in this chapter would offer a more comprehensive picture, with annoyance being *one* of the indexes used. Furthermore, in developing a land use inventory of noise exposure in an airport community, aircraft should not be the only noise source considered, even though it may be predominant. Major thoroughfares and industrial activities, often associated with airport communities, are additional factors that could easily affect land use compatibility.

Comprehensive Planning

A major step toward solving or mitigating the airport noise problem is to change the present scale of comprehensive planning. Planning occurs—but not on a regional level. It is fragmented. The airport has developed a master plan as a guide for future development, indepen-

Aircraft Noise and Residential Exposure: Tinicum Township   TABLE 7.3.

Aircraft operation	Noise index		Residential exposure
	Composite noise rating (CNR)	Noise exposure forecast (NEF)	Percentage of residences affected*
Takeoff (Present runway)			
Daytime	115	>40	60%
	100	>30	100%
Nighttime	115	>40	80%
	100	>30	100%
Takeoff (Future runway)			
Daytime	115	>40	20%
	100	>30	100%
Nighttime	115	>40	60%
	100	>30	100%
Landing (Present runway)			
Daytime	115	>40	10%
	100	>30	60%
Nighttime	115	>40	20%
	100	>30	60%
Landing (Future runway)			
Daytime	115	>40	0%
	100	>30	20%
Nighttime	115	>30	0%
	100	>30	30%

*Percentage estimates rounded off to nearest 10%.

dent of the adjoining Township. In that plan, *no mention is made of the community.*[32] The question of airport compatibility is conspicuously absent. Delaware County, which is responsible for planning Tinicum Township, neglects to mention the airport and the existing compatibility problem, even though most of the runway lies in the county.[33] Planning activities in the City of Philadelphia and Delaware County take place independently of one another, because the two areas represent different political jurisdictions. Until their activities are combined, or at least coordinated, the airport condition will remain the same or worsen. A regional planning body already exists, the Delaware Valley

Regional Planning Commission. This group could be a useful arbiter, or even become responsible for the metropolitan area planning of this airport.

Zoning

Once planning guidelines for future growth of airports and their neighboring communities are considered jointly, zoning ordinances should be enacted to minimize the airport noise problem. Compatibility, as we have pointed out before, is a two-way street; its most important objective is environmental health protection. To assure such protection will require the strict enforcement of any airport zoning ordinance once adopted. Differences in zoning which involve residential encroachment into the path of more intense aircraft noise, and unplanned airport expansion into residential areas, should not be permitted.

The basis for the planning and subsequent zoning of this area should be noise exposure. Land uses closest to the airport noise should contain only indoor nonresidential activities, low density outside activities (for example, sanitary landfill, sewage or water treatment), and ground transportation systems. Sound insulation requirements would be established for the various noise zones, depending upon location and type of land use. People living in areas rezoned as nonresidential because of noise intensity would have their properties designated noncompatible. Those owners would not be allowed to sell their homes for residential purposes. However, an equitable price would be arranged, so that these homeowners would receive a fair market value for their property.

Community Representation

The fact that people adjacent to the airport have no voice in the development of the airport has heightened their response to the noise it produces. A meaningful dialogue between airport officials and the community is necessary to remove the community's apprehension. The airport operators also cherish a stereotype of local residents as troublemakers unduly sensitive to noise. Polarization is developing because the Township has no responsive outlet to which it can air complaints or offer solutions.

Associated with the metropolitan area planning of the airport should be the establishment of an airport noise control council. The council would include both municipal and local community representatives. Once an airport noise program was established, the council could monitor it, handling any inquiries, complaints, or violations with some professional guidance. They could also be an educational agency, explaining problems, presenting possible solutions, and reporting concrete steps taken. Finally the council's responsibility could include establishing guidelines for the allocation of funds to compensate homeowners in noncompatible areas, as well as to make actual awards.

The most feasible and practical method to deal with noise *where it is perceived* is by sound insulation. This solution was first tried by the London Airport Authority to reduce noise in homes adjoining London's Heathrow Airport. A similar program in this country was begun in 1967, when the Los Angeles City Council approved the request by the Airport Authority to issue $75,000,000 worth of revenue bonds to

(a) acquire houses as part of a runway and terminal expansion program
(b) begin a modern soundproofing program to improve the habitability of residential, school, and church structures.[34]

Significant reductions of intrusive noise can be obtained, depending upon the type of dwelling, degree of sound insulating, and so forth. A greater problem is to determine an equitable distribution of the costs. In London, for example, although £2,000,000 have been set aside, only £8,500 have been spent, because each homeowner must pay one third of the cost, or £100.[35,36] (Starting with a minimum, a sliding payment scale could be applied, depending on the number of years in a residence and the level of noise exposure.)

Besides the financial burden, there are two basic limitations to sound insulation. In warm climates, not only will sound insulation be necessary, but also air conditioning, since a noise-insulated house must be closed all seasons of the year. This raises the cost of sound-insulation considerably. Second, the homeowner is restricted in the use of his property. For all practical purposes, his enjoyment of the outdoors is curtailed. Only the house itself is a refuge from community-generated noise. The cost of sound-insulation is not inexpensive, as found in the demonstration project sponsored by the Los Angeles Department of Airports. Depending upon the extent of soundproofing, costs ranged from $3,210 to $12,550 for houses valued between $20–30,000 [37] (see Table 7.4). Acoustical modifications incorporated in Stage One were judged to be unacceptable to the homeowners. Stage Two and Stage Three modifications reduced the noise sufficiently for the homeowner but the costs increased appreciably. An extensive residential sound insulation program should not be undertaken when such a program will produce marginal results.

RECEIVER/OBJECT

Home Soundproofing Study: Los Angeles   TABLE 7.4.

	Stage One	Stage Two	Stage Three
Cost	$3,210	$5,620	$12,550
Sound level reduction	25 dB	35 dB	50 dB

Residential relocation may be necessary under certain noise conditions. Consequently efforts should be made to provide an orderly and smooth relocation process for those inhabitants involved. Several residential areas around the Philadelphia International Airport should be relocated. Even if home insulation techniques could be successfully applied, these people in Tinicum would not have regained normal use of their property. Site acquisition costs as in the case of acoustical modification may be sizable. The Los Angeles airport commissioners agreed to purchase 1,994 homes near the take-off and approach paths at Los Angeles International Airport. When this project is completed in 1973 the city will have spent, with interest, nearly $300 million.[38]

In Oberbolheim, Germany, the West German government gave careful consideration to the airport noise problem. The government ultimately decided to relocate the entire town of Oberbolheim, a population of 225 residents. A planned community of Neu Oberbolheim was developed and subsequently inhabited.[39]

Current practices and proposed airport improvements will not effectively reduce the noise problem. There are many possible ways to reduce the level of airport noise to which the surrounding areas are exposed. A comprehensive noise management program, including source, path, and object controls is necessary. Such a program will require a new governmental framework for effecting these controls, including municipal as well as community representation. Environmental planning should be the basis of this new framework. A new town in France, Vaudreuil, is being planned according to environmental health criteria including the control of noise.[40] What is at stake is the environmental health protection of the population. Philadelphia, as well as many other noise-inflicted cities, cannot afford to disregard this pollutant any longer.

O. If *yes* what, and if *no* why not?

   1. _____ 2. _____ 3. _____

   _____ _____ _____

P. Would you be willing to bear part of the cost to reduce the noise in the area where you live? 1. Yes _____ 2. No _____

Q. If no, why not?

   1. _____ 2. _____ 3. _____

R. If yes, what method or methods of raising the money would you suggest?

   1. Taxes _____ 2. Voluntary Contribution _____ 3. Other _____

S. What amount would you be willing to spend for noise control?

   1. $1–5 _____ 3. $10–15 _____ 5. _____

   2. $5–10 _____ 4. $15–20 _____

T. If your local noise problem cannot be solved, do you believe you should receive some financial compensation?

   1. Yes _____ 2. No _____

   What amount should you receive?

   1. Under $100 _____ 3. $500–1,000 _____

   2. $100–500 _____ 4. $1,000–2,000 _____

U. Who should pay you this compensation? _____

III. DEMOGRAPHY

A. Address _____ _____

   House Number   Street   Municipality

B. Respondent: Head of household _____ 2. Spouse _____

              3. Other _____

C. Sex of Respondent: 1. Male _____ 2. Female _____

D. Date of Birth _____

E. Race: 1. White _____ 2. Negro _____ 3. Other _____

F. How many persons presently live at this address?: 1. How many of these persons are six years and under _____

   2. Between 7–15 yrs. _____ 3. Between 16–21 yrs. _____

G. How many rooms does your home have?: 1. _____

   How many of these are bedrooms?   2. _____

H. Type of Dwelling Unit?

   1. single family _____   4. Flat   _____

   2. Semi-detached _____   5. Apartment _____

   3. Row house _____

I. Do you own? 1. _____ (or) Rent 2. _____

J. Does your dwelling have:

              Yes No   Yes No

   1. Central   3. Window fans _____

     Air-conditioning _____

   2. Window air-   4. Storm Windows _____

     condition units _____

K. How long have you lived at your present address?   1. _____
   How many years have you lived in Philadelphia   2. _____
L. Within the past 10 years how many times have you moved?
   1. _____
M. How many of these moves were: 1. Within your present neighbor-
   hood _____   2. Other parts of Philadelphia _____   3. Out-
   side the city _____
N. Where was your last residence?   1. Address _____
   2. City/State _____
O. Do you believe the move to your present residence was an improve-
   ment?   1. Yes _____   2. No _____
P. Do you intend to move?   1. Yes _____   2. No _____
   3. Comments _____
Q. During most of your years, up to the age of 18, where was your home?

Years	Municipality	State	Population	Urban	Suburb.	Rural
1.						
2.						
3.						

R. In comparison with where you presently live, would you consider
   that _____ home:
   1. Much quieter _____   2. Quieter _____   3. Same _____
   4. Noisier _____   5. Much noisier _____ .
S. Are you currently employed?   1. Yes _____   2. No _____
   (Retired _____   Disabled _____)
T. (a) How many hours do you work a week?   1. _____
   (b) How many months have you worked in the past year?
       1. Months _____
U. Where is your place of work?   1. Male _____
                                  2. Female _____
V. (a) What are the hours of your job (s)?
       1. _____ a.m.   to _____ p.m.
       2. _____ a.m.   to _____ p.m.
   (b) Describe the type of job you have? _____
W. How would you rate your work environment in regard to noise?
   1. Very quiet _____   2. Fairly quiet _____   3. Slightly
   noisy _____   4. Noisy _____   5. Very noisy _____
X. What is the highest grade of school completed?   1. _____
Y. How many hours a day do you have your radio and/or television on?
   1. _____
Z. What is the total family income supporting this residence?
   1. Under $5,000 _____   3. $10–15,000 _____   5. No reply _____
   2. $5–9,000 _____   4. $15,000 + _____

National Advertising of Consumer Products

A. *Automobile Tires*

COMPANY	PRODUCT	ADVERTISING COPY	PERIODICAL
Atlas	Snow tire Atlas Weathergard	"If you made a great snow tire, would you keep it quiet? Atlas did." "Now Atlas gives you the quiet-running giant of a winter tire that fills the bill completely."	*Newsweek* December 2, 1968
		"Atlas Weathergard runs quiet, too. On dry roads, you don't get as much annoying whine as from ordinary snow tires. Since up to 80% of winter driving is on snow-free roads, you'll find our quiet-running Atlas Weathergard will let you hear yourself think."	*Newsweek* October 27, 1969
		"Atlas Weathergard tires run quiet, too. There's no high pitched whine on dry roads."	*Life* October 30, 1970
Delta Tire Corporation	Delta 140 Super Premium QT	"Quiet At Last! A Tire That Gives You Total Safety . . . Without Thump." "No annoying noise regardless of the weather."	*Sports Illustrated* April 15, 1968

A. *Automobile Tires (Cont'd.)*

COMPANY	PRODUCT	ADVERTISING COPY	PERIODICAL
	Delta Duraglas	"A smooth, quiet ride."	*Newsweek* June 9, 1968
		"Less stress, less wear, less noise and up to twice the mileage, too."	*Newsweek* June 23, 1968
	Delta Premium 178 Belted	"Assures a smooth, quiet ride . . ."	*Life* June 19, 1970
Dunlop	Dunlop Snow Tire:	"On a Dunlop tread designed to carry you without noise or vibration."	*Newsweek* October 27, 1969
	Dunlop Silent Traction	"Take a summary ride through winter" . . . "They'll tread smoothly and quietly on all the clear winter roads ahead."	*Newsweek* November 9, 1970
Firestone Tire Company	Firestone Town and Country wide oval tire	"Town and Country Wide Oval tires give a smooth, quiet ride."	*Reader's Digest* November, 1969
	Firestone Transport 110	"For example, there's the Transport 110, an all-wheel tire from Firestone with quiet running and long mileage built in."	*Popular Science* August, 1970
	Firestone Town & Country	"This new Town & Country has a special new flat contour tread design for quieter ride . . ."	*Newsweek* October 26, 1970
	500 Wide Oval	". . . a whisper-smooth ride."	*Popular Science* March, 1971
Kelly Springfield	Kelly Springfield Wintermark G/P	"New deep-cleat tread for great mud/snow and ice; quiet ride on dry pavement."	*Life* December 19, 1969
Pirelli Tire Corporation	Pirelli Cinturato CN72—The Flat Tire	"What's more, The Flat Tire rides smoothly and silently on any surface. At any speed."	*Newsweek* June 2, 1969
Sears Roebuck Company	Sears Super-wide Fiber Glass Snow Tire	"A good snow tire should be seen and not heard."	*Newsweek* November 20, 1967
		"The Sears Silent Guard Sealant Tire."	*Sports Illustrated* April 14, 1969
Seiberling Tire and Rubber Company	Seiberling Four Seasons	"At last. A snow tire that's quiet when it runs out of snow. Seiberling engineered a new idea into the Four Seasons, and now the days of whine and road hum are gone.	*Newsweek* October 27, 1969

A. *Automobile Tires (Cont'd.)*

COMPANY	PRODUCT	ADVERTISING COPY	PERIODICAL
		". . . It also makes it sound like one. Quiet so the Four Seasons is quiet and it pulls. . . . That's for strength, durability, control and an even quieter ride. No one will know you're driving on snow tires. And they won't hear about it unless you tell them. See your Seiberling dealer. And pull through the winter quietly, safely."	
Uniroyal	Uniroyal Masters	"(We'd like to mention that although our rear tires can function as snow tires, they're not noisy like snow tires. That's because the deep-lug thread is on the inside of the tire, so that the noise factor is dissipated underneath the car.)"	*Newsweek* May 26, 1969

B. *Household Appliances*

COMPANY	PRODUCT	ADVERTISING COPY	PERIODICAL
General Electric	Dishwasher	"Each one also has three jet streams that thoroughly but quietly wipe out dirt in a cross fire of water." "The new GE dishwasher has four jet streams. When they catch dirt in a quiet crossfire, they don't let up until every last speck is wiped out."	*Life* October 11, 1968
Hoover Company	Vacuum cleaner	"Although it's whisper quiet, it's powered with a big full horsepower motor."	*Life* August 15, 1967
Kitchen Aid Hobart Mfg. Co.	Kitchen Aid disposer	"It goes about its work so quietly you'll hardly know its around."	*Better Homes & Gardens* December, 1970
Pfaff American Sales	Pfaff 78 Sewing Machine	"Silencing is improved, vibration dampened with a new isolation system."	*Good Housekeeping* April 1, 1969
Montgomery Ward	Signature dishwasher, dryer, and refrigerator	"Silence is golden" . . . "you'll hear a beautiful nothing from our Signature appliances. No rattles. No whines. No malfunctioning moans. Because this year like every year we've built in the same quiet dependability." "It's as essential to us as our appliance features could be to you."	*Life* July 10, 1970

B. Household Appliances (Cont'd.)

COMPANY	PRODUCT	ADVERTISING COPY	PERIODICAL
		"For example. The ice maker in our Signature Refrigerator is so quiet you won't hear as much as a hum from it." "And it will wash them with the minimum of noise and vibration." "Even our Signature Dishwasher has a cushioned motor to insure quiet."	
Speed Queen	Clothes dryer	"Then there's the quiet blue gas flame that runs it."	*Ladies Home Journal* October, 1970
Waste King Universal	Garbage disposer	"Hush quiet and dependable."	*Good Housekeeping* December, 1970

C. Air Conditioners

Carrier Air Conditioning	Central air conditioning	"The world is full of things you can hear. But, one man's sound is another man's noise. No matter which it is to you, when something important is going on Carrier air conditioning doesn't interfere. Quiet operation is one of the reasons more people put their confidence in Carrier than in any other make."	*Newsweek* 1966
		"Buy quiet. Air conditioning your home is a sound way to shut out noises from outside. Carrier goes one step better. Makes the air conditioner itself quiet—by design. Like high-density insulation. Cushion mounted fans and compressors. Acoustically engineered throughout."	*Life* April 21, 1967
		"When all the other noises have gone to bed, you'll know how quiet Carrier is. During the day it doesn't matter what kind of air conditioning keeps your home comfortable. But when things quiet down Carrier comes into its own. With extra sound insulation. Cushion mountings. Advanced acoustical engineering. Years from now,	*Life* March, 1968

C. Air Conditioners (Cont'd.)

COMPANY	PRODUCT	ADVERTISING COPY	PERIODICAL
		when other noises have grown up, you'll become even more aware of the value of Carrier . . ."	
		"Being round, its more attractive. Compact. Efficient. Quiet. Very considerate of your neighbors, your shrubbery." "Heat and sound are blown straight up. Away from shrubs. Away from neighbors."	*Newsweek* May 19, 1969
		"More coil surface is exposed to outside air for quieter, more efficient cooling. The fan, a special design, moves more air quietly . . . It blows the heat and sound up, away from your flowers and neighbors. The hermetic compressor is nestled deep inside the unit for more quiet."	*Life* March 13, 1970
		"We nestled our hermetic compressor deep inside to absorb its murmur. Used dual mufflers for a neighborly kind of quiet."	*Life* April 24, 1970
		"The Round One has built-in silencers for extra quiet." "Its slower turning fan moves more air than most, blowing heat and sound up, away from neighbors. Yet it exposes more coil surface to outside air for quiet more efficient cooling."	*Life* May 1, 1970
Friedrich	Window air conditioning	"To the discriminating few who want the ultimate in quietness and quality." "Friedrich room air conditioners are so quiet." "Friedrich is soundly constructed to keep sound down."	*New York Times* August 3, 1970
Frigidaire	Window air conditioning: Prestige Model AEP-8MN	"It roars like a mouse. We're talking about the pindrop quiet air conditioner from Frigidaire . . . the one that works with just a soft, gentle murmur. It runs so quietly you can almost hear yourself think. To sum up . . . if you ever spent the night with a noisy air conditioner, you'll appreciate what a blessing pindrop quiet cooling can be."	*Life* June 7, 1968

C. *Air Conditioners (Cont'd.)*

COMPANY	PRODUCT	ADVERTISING COPY	PERIODICAL
		"This Frigidaire air conditioner does some-thing extra. It keeps quiet! Cool should be felt . . . not heard." "This Frigidaire air conditioner model has lots of insides specifically designed to over-come annoying air conditioner noise prob-lems. It's pin-drop quiet. Buy one and rediscover the joys of summer conver-sation."	*Life* May 30, 1969
Frigiking Division of Cummins Engine Co., Inc.	Automobile air condi-tioning	"And Frigiking's engineering heritage shows itself. To cool your car quickly, quietly as only Frigiking can."	*Look* June 10, 1969
General Electric	Window air conditioning: Fashionaire	"The quiet-cooling new GE Fashion-aire . . ." "There's slumber speed that doesn't snore like many air conditioners: our new low level fan speed that's extra quiet for light sleepers."	*Reader's Digest* June, 1969
	Central air conditioning: The Executive	"Most of the time it operates on the low fan speed. And very quietly." "The hot air removed from your home is blown upwards so it doesn't scorch nearby plantings or grass. Much quieter for your neighbors, too!"	*Newsweek* July, 1969
	Central air conditioning	"So it's reliable, our compressor . . . An added plus, your neighbors will love how nice and quiet it is."	*Newsweek* September 15, 1969
	Window air conditioning	"Not only does it adjust the fan speed to fit the situation, by doing so, it automati-cally adjusts for cooling, for humidity con-trol and for quietness."	*Life* July 4, 1970
General Motors	Automobile air condi-tioning	"Air conditioning to make your driving cool, quiet, pollen and dust free."	*Life* November 15, 1968
Lennox Industries, Inc.	Central air conditioning	"You live and work daily with Lennox all around you. Quietly, efficiently, comfort-	*Newsweek* September, 1968

C. Air Conditioners (Cont'd.)

Company	Product	Advertising Copy	Periodical
		ably cooling, heating and ventilating so many buildings in your town."	
Payne Company	Central air conditioning	"Payne's all-new Remote air conditioner will snug right up against the house and hide behind the roses and run without rattling and without waking up the neighbors and withstand nasty weather and send the sound up and away and help keep the roses from wilting and the dog from being so grouchy . . ."	*Newsweek* June 9, 1969
Westinghouse	Window air conditioning	"We're creating new ways to learn in school while we give you a new way to cool your home quietly." "Westinghouse helps in your home by giving you a new kind of room air conditioner designed to operate more quietly."	*Life* May 16, 1969
		". . . Westinghouse is creating better living conditions at home by building the quietest room air conditioner you can buy."	*Life* May 22, 1970
Whirlpool Corporation	Window air conditioning	"How come these two Whirlpool air conditioners with the same cooling power don't cost the same? That's easy—the $189.95 model gives you four extra features. A decorator panel that muffles noise."	*Look* June 10, 1969
	Model AVM-090-2 (9000 BTU) and AVL-210-3 (2100 BTU)	"And both have these outstanding features. A decorator panel that not only looks good but muffles sound as well."	*Newsweek* January 19, 1970
York Air Conditioning Division of Borg-Warner Corp.	Central air conditioning	"York's new comfort conditioning system is low, compact, quiet."	*Life* May, 1968
		"York's new whole-house air conditioner runs so smoothly, so quietly, your neighbors will never *hear* how comfortable you are!"	*Time* May 10, 1968

C. Air Conditioners (Cont'd.)

COMPANY	PRODUCT	ADVERTISING COPY	PERIODICAL
		"Now you can enjoy whole-house comfort conditioning without making a big noise about it! York has found a quiet way to air condition your home . . . moving parts are isolated in a sound deadening chamber. The powerful motor and fan run slowly, quietly."	
		"Replace your old air conditioner with York. It'll comfort you—and won't make a big noise about it." ". . . with a modern York unit that gives you dependable cooling, quietly. York has engineered the noise out of air conditioning —left the comfort in. Moving parts are isolated. Motor and fan run slowly, quietly, assuring years of trouble-free operation."	*Newsweek* May 26, 1969
		"Only 18 inches high. Quiet motor and fan reduce sound."	*Life* June 7, 1968
		"And it has a quiet motor and fan. So you and your neighbors won't be bothered by a lot of noise."	*Time* March, 1971

D. Power Equipment

COMPANY	PRODUCT	ADVERTISING COPY	PERIODICAL
Ariens Company	Lawnmower: The Peacemaker	"The Peacemaker. Designed to keep peace in the family and the neighborhood."	*House and Garden* April, 1968
Black & Decker	Electric Mower: Single-blade rotary	"Cuts carpet-smooth, and runs quietly."	*Popular Science* March, 1971
Ingersoll-Rand	Air Compressor: diesel-powered portable	"Dig Quietly We Must! The whisperized compressor is here. You'll have to believe it. A 900-cfm diesel-powered portable air compressor that's so quiet you could dictate a letter right beside it, while it's running at full load." "Ingersoll-Rand's new Whisperized SPIRO-	*Forbes* August 15, 1969

D. *Power Equipment (Cont'd.)*

COMPANY	PRODUCT	ADVERTISING COPY	PERIODICAL
		FLO compressor makes only one hundredth as much noise as a conventional unit of the same size. Its full-load sound rating of 85 decibels at a distance of 3 feet is well within limits specified in a recent New York legislative proposal on noise abatement." "It's the answer to growing demands in cities, residential areas, hospital zones, business districts—wherever noise must be minimized."	
Mowbot, Inc.	Lawnmower: electric	"Rechargeable electric power makes Mowbot a good neighbor. No noise. No fumes."	*House and Garden* May, 1969
	Lawnmower: electric	"Introducing quiet new single-blade electric lawnmowers." "A power mower that keeps the peace! Almost silent."	*Better Homes & Gardens* April, 1969
Stihl America Inc.	Chainsaw: Stihl 030AV	"a super silent muffler."	*Popular Science* February, 1971

E. *Vehicles*

COMPANY	PRODUCT	ADVERTISING COPY	PERIODICAL
American Honda Motor Co., Inc.	Motorcycle: Honda Mini Trail	. . . "Honda has built in a USDA approved spark arrestor/muffler and a fully enclosed exhaust system to help keep the wilderness just the way you found it. Quiet and green."	*Look* August 25, 1970
American Motors	Automobile: Rambler American	"We put all the good things that people would expect to find in an American car into our Rambler 2-door sedan. Horsepower. Room. A quiet, steady ride."	*Newsweek* February 10, 1969
Artic Enterprises, Inc.	Bike: Artic Cat	"Quick, quiet, easy riding and tough . . ."	*Popular Science* April, 1971
British Leyland Motors, Inc.	Automobile: Rover 2000	"The Rover is quiet inside save for the engine's reassuring purr. Thick wall-to-wall carpeting absorbs lots of noise."	*Esquire* May, 1969

E. Vehicles (Cont'd.)

COMPANY	PRODUCT	ADVERTISING COPY	PERIODICAL
British Motors Corporation	Automobile: MGB	"Even the MGB's aerodynamic, wind-cheating body is functional. Aside from looking attractive, it means quieter running with less wind noise, and greater economy."	*Time* January 21, 1968
Chrysler Motors Corp.	Automobile: Chrysler	"Inside your next car, a cool, quiet room of curved glass and tempered steel."	*Life* October 11, 1968
		"A thick layer of insulation shuts out world noises and leaves your cockpit nearly soundless."	*Life* October 18, 1968
		"You can get quiet rides with other cars, but with Chrysler cars you get the ideal combination of quietness, stability and control . . . Your next car: 1970 Chrysler with Torsion-Quiet Ride."	*Newsweek* September 29, 1969
		"Less noise; less rust; more performance. Unibody is quieter, too. There is far less flexing than you'll find in a separate body and frame—more protection against opened joints, popped welds—flaws that can admit water and salt and promote rust. Also, there is less vibration—less chance for rattles and squeaks. (Keep in mind, too, that welds can't rattle; nuts and bolts can.)"	*Life* September 11, 1970
		"The quiet car gets quieter for 1970. New rubber body mounts, new suspension system isolators and 25 sq. ft. more of sound insulation. Chrysler's new Sound Isolation System." "Chrysler's unibody construction. 5,000 individual welds produce a unit of unusual strength . . . silence . . . and durability."	*Ebony* October, 1970
		"Torsion-Quiet Ride. No one else offers it." "The solid Unibody is welded together. There are no body bolts to work free and rattle after a year or so of driving."	*Newsweek* October 19, 1970

E. *Vehicles (Cont'd.)*

COMPANY	PRODUCT	ADVERTISING COPY	PERIODICAL
	Automobile: Ford Thunderbird	"Its special ride has always meant smoothness, comfort, quiet."	*Newsweek* September 28, 1970
	Automobile: Lincoln Continental	"A ride as cathedral quiet . . ."	*Newsweek* March 15, 1971
	Automobile: Mercury Marquis	"The Marquis is also one of the quietest riding cars ever built. Thirty important areas of the car have been carefully insulated with sound-deadening material."	*Newsweek* September 29, 1969
		"Extra sound insulation to hush wind and road noise."	*Business Week* October 17, 1970
		"Extra insulation is used in 30 areas of the car to hush wind and road noise."	*Newsweek* November 16, 1970
	Truck: Ford Diesel	"Ford's V-Series Diesels (V-150, V-175, V-200, V-225) are quiet, virtually smoke-free and engineered for maximum reliability in short-haul construction operations."	*Engineering News-Record* June 12, 1969
	Truck: Ford Econoline Van	"Ford's engine cover is heavily insulated to seal out heat, noise."	*Newsweek* January 12, 1970
	Truck: Ford Louisville Line	"Carefully insulated against noise, vibration, temperature extremes."	*Newsweek* January 12, 1970
	Truck: Ford	"Levelest, quietest ride of them all . . ." "Independent tests proved Ford is quieter than all other leading pickup makes."	*Newsweek* October 19, 1970
	Truck: Ford Louisville Line	"Ford offers the most advanced cab in the industry. Sound, temperature, and vibration insulated."	*Newsweek* November 16, 1970
	Truck: Ford	"Welded, instead of bolted, pickup box stays strong, resists working loose, means greater quietness . . ."	*Reader's Digest* January, 1971

E. Vehicles (Cont'd.)

Company	Product	Advertising Copy	Periodical
General Motors Corporation	Automobile: Cadillac	"The big 472 V-8 engine smoothly and quietly delivers a responsiveness that's astonishing for a car of such magnificence."	*New Yorker* October 11, 1969
		"The quietness of operation will impress even longtime Cadillac owners."	*Sports Illustrated* December 14, 1970
		". . . the elegance of interior appointments, the comfort of the ride and the almost un-believable smoothness and quietness of their performance."	*Newsweek* January 18, 1971
		". . . with a smoothness and quiet never before known to the American luxury automobile."	*Reader's Digest* February, 1971
	Automobile: Buick Le Sabre	". . . silent Body by Fisher."	*New Yorker* April 18, 1970
	Chevrolet Chevelle	"Refined rear suspension for a smoother, quieter ride."	*Life* May 29, 1970
	Automobile: Chevrolet	"Chevrolet put it all together. Solid gentle-manly comfort without bomblast. Sailplane silence."	*Forbes* October 1, 1969
	Automobile: Chevrolet Monte Carlo	"And so quietly beautiful." "With Monte Carlo, we did more to nullify noise than make all those welds. And posi-tion plump rubber biscuits at critical body mount areas. We made silence a science. Every car ever made has its own peculiar acoustics. It has certain "holes" that admit sound. And certain noise paths that trans-mit and amplify that sound." "What we did was track them down. Then dam them up with much more blanket and spray insulation than is customary on a car this size." "It all adds up to a Monte Carlo we can only describe as sailplane silent."	*Time* October 17, 1969

E. Vehicles (Cont'd.)

COMPANY	PRODUCT	ADVERTISING COPY	PERIODICAL
	Chevrolet Impala	"They could hush up the sounds of old age with something like Impala's long life exhaust system."	*Life* March 20, 1970
		"And our new suspension. Smoother. Quieter. But you'd like all the peace of mind and quiet you can get, too." "With its new roof, which puts two welded steel roof-panels over your head. Stronger. Quieter."	*Newsweek* March 22, 1971
	Automobile: Chevrolet Monte Carlo	"And take a listen, too. There's extra thick insulation for added quiet."	*Life* October, 1969
		"We made it stronger. And we made it quieter by putting a double layer of steel in the roof."	*Life* October 2, 1970
	Chevrolet Vega	"Yet it hums along with a degree of quietness that is all too unusual in little cars." ". . . Nor is it as noisy as an engine that's turning faster."	*Newsweek* January 25, 1971
		"Wait until you hit the highway and feel how smooth and steady and quiet a Vega runs at highway speeds . . ."	*Newsweek* March 8, 1971
	Oldsmobile	"The whisper-quiet Flo-Thru Ventilation now features both upper and lower level vents."	*Life* December 18, 1970
		"You enjoy the same kind of quiet on the road that you enjoy in your own library."	*Newsweek* February 1, 1971
	Opel	"And it has new hydraulic valve lifters to make it run quietly and eliminate the need for adjustments."	*Sports Illustrated* December 14, 1970
	Pontiac Lemans	"The vent windows are gone altogether— for less wind noise and cleaner lines."	*Look* January 12, 1971

E. Vehicles (Cont'd.)

COMPANY	PRODUCT	ADVERTISING COPY	PERIODICAL
	Pontiac Grand Ville	"Pontiac's smoothest, quietest ride ever."	*Time* January 4, 1971
	Truck: Chevrolet	"Special insulation and body mounts to act down on noise and vibration.	*Life* November 15, 1968
	Truck: GMC	"Unsurpassed acoustical and thermal insulation provides maximum quiet, maximum comfort all year around."	*Newsweek* November 25, 1968
	Truck: Chevrolet	"It has double walls in the cab and load box, with insulation in the roof for added quiet."	*Reader's Digest* December, 1969
International Harvester Company	Truck: International Trucks	"And the dressed-up cabs of our big TRANSTAR—a new world of quiet comfort anyone would enjoy."	*Newsweek* January 20, 1969
		"Stronger, more economical, easier to service, quieter . . ."	*Newsweek* March 24, 1969
Massey-Ferguson, Inc.	Tractor: Massey-Ferguson, Inc.	"For an extra measure of safety and quiet order the new MF cab."	*Progressive Farmer* April, 1969
Renault	Automobile: Renault 16	"And the quietness of the Renault 16 comes only in cars costing thousands more."	*Newsweek* May 11, 1970
Rolls-Royce Ltd.	Automobile: Rolls-Royce Silver Shadow	"The Silver Shadow's silence, even at 100 miles per hour, adds to your peace of mind."	*New Yorker* June 14, 1969
		"Rolls-Royce engineers are fanatic about engine silence. They weigh and balance all engine parts before assembly, run each engine for hours on a testbed for further refinement, then muffle its exhaust with four stainless steel silencers. Even while crusing at 100 miles per hour, you can chat with your passengers without raising your voice."	*New Yorker* April 18, 1970
Toyota Motor	Automobile:	"A ride that's soft, and library quiet."	*Look*

E. Vehicles (Cont'd.)

COMPANY	PRODUCT	ADVERTISING COPY	PERIODICAL
	Toyota Corona		December 18, 1968
		"And a 60 hp engine that delivers about 30 miles to the gallon, along with the fastest, quietest ride in the low price field."	*Look* June 10, 1969
		"And the solid, quiet ride is a complete surprise."	*Newsweek* June 23, 1969
		"And a newly engineered suspension system, front and back, for a smoother ride. And also to deaden sound."	*Life* September 11, 1970
	Automobile: Toyota Corolla	"With unitized construction for a safer, quieter ride."	*Life* November 21, 1969
		"And provides a soothingly quiet interior —air conditioned if you like."	*Look* February 24, 1970
		"And reach speeds of 87 miles per hour. Quietly."	*Life* March 20, 1970
		"With undercoating to prevent rust, corrosion and noise. With unit construction and a lined trunk to prevent rattles and squeaks."	*Life* October 30, 1970
	Automobile: Toyota Mark II	"Now you can't even hear the engine" "It idles so quietly there are times you'll swear it died." ". . . And economical without being noisy. By designing the engine with an overhead cam, we cut down the number of moving parts. Which reduces both wear and noise. By making the fan out of nylon we made it lighter and quieter."	*Life* January 22, 1971
Volvo	Automobile Volvo 144	"We replaced our old radiator fan with a new one that's automatically limited to 3000 rpm. Its quieter and soaks up less power."	*Newsweek* February 3, 1969

F. Aircraft

Company	Product	Advertising Copy	Periodical
Beechcraft	Beechcraft King Air	"It's the first of the new sophisticated business airplanes to combine high cruising speeds, big payload, the comfort of a pressurized cabin at over-the-weather altitudes, plus the smoothness and whisper-quiet of turbine power . . . without sacrificing operating economy at low altitudes, or the added ability to land and take off from small, short airports."	*Time* June 20, 1969
Boeing Company	Boeing 747	"Now undergoing the most extensive test program in the history of commercial aviation, the 747 has proved extremely smooth, quiet and steady . . ."	*Newsweek* October 27, 1969
		"When your 747 flight takes off, four of the world's most powerful but quiet commercial jet engines will lift you gracefully into the air."	*Reader's Digest* January, 1970
British Overseas Airways Corp.	BOAC: VC10	"It started a quiet revolution." "Once upon a time, flying a big transatlantic jet was rather a noisy experience. Then, about three years ago, BOAC introduced a transatlantic jet with Rolls Royce engines back by the tail instead of on the wings. It left all the usual cabin noise behind. So the people inside could hear themselves think. The name of this jet is the VC10, and it has quietly become the most in demand airplane from here to London. Peace and quiet aren't the only things it has to offer."	*Newsweek* February 13, 1967
		"Our cabins became as quiet as an English meadow on Sunday afternoon. Because the noise was all behind us."	*Newsweek* September 7, 1967
		"Offers a virtually noisefree ride."	*Newsweek* November 7, 1968

F. *Aircraft (Cont'd.)*

COMPANY	PRODUCT	ADVERTISING COPY	PERIODICAL
		"A cabin that's virtually noise and vibration free, thanks to the VC10's rear-mounted Rolls-Royce engines."	*Newsweek* January 20, 1969
		"We want you to leave home quietly. And peacefully. We want you to fly on the quietest, most advanced commercial jet in the world. The BOAC VC10. The only jet with four Rolls-Royce engines mounted way back at the tail to leave all the roar outside instead of in."	*New Yorker* May 17, 1969
		"Its rear-mounted engines let the cabin remain as quiet as a lagoon."	*New Yorker* September 13, 1969
		"And if you're flying on one of our remarkable VC10's, you'll enjoy the quietest cabin in the air."	*New Yorker* October 25, 1969
		". . . four powerful rear-mounted engines to leave all the noise behind."	*Time* November 16, 1970
Eastern Airlines	Whisper Jet	"Whisper Jet"	
	Air Freight	"The fantom warehouse is an Eastern Air Freight container that holds about 4500 lbs. of cargo. Eight of them fit into one of our Whisperjet freighters."	*Time* June 20, 1969
General Electric	General Electric DC-10	"These engines are among the most powerful of their kind in the world. But they're also exceptionally quiet and economical."	*Time* June 13, 1969
		"Jet noise is a different problem. But we— and the Federal Government—are tackling this one, too. NASA has called on General Electric to help solve the problem for the aviation industry." "We have already succeeded in making the DC-10 engine quieter than the engines now	*New Yorker* August 22, 1970

F. Aircraft (Cont'd.)

COMPANY	PRODUCT	ADVERTISING COPY	PERIODICAL
		powering most of today's jet planes. Yet this GE engine is nearly three times as powerful."	
Hawker Siddeley International, Inc.	Hawker Siddeley DH125	"It lets you conduct meetings in the plane, head to head, without acting like a flock of screaming eagles. Those Rolls-Royce engines are so quiet you can hear yourself think."	*Forbes* October 15, 1969
Lockheed Aircraft Company	Lockheed 1011	"Quieter power. Three mighty Rolls-Royce high-bypass turbofan engines will make the quietest 'sh-h-h' in the sky. Take-off and landing noise will be far below present jet levels, making this airliner a quieter, better neighbor for people on the ground as well as those in the air. In 1971, the low-noise Lockheed 1011 will start flying for many great airlines."	*Newsweek* August 24, 1968
		"A trio of extraquiet Rolls-Royce engines will power it to shorter take-offs. So it will fly from airports now off limits to many airlines."	*Time* May 30, 1969
		"And Rolls-Royce's new three-shaft engines are the smoothest and quietest that ever flew."	*Time* January 4, 1971
McDonnell-Douglas	Douglas DC-9	"This quick and quiet jetliner makes it possible for airlines to schedule more flights to meet the needs of today's businessman."	*Time* May 10, 1968
		"We also placed its twin fanjets at the rear, to give you smoother, quieter flights."	*Time* May 16, 1969
Piedmont Airlines	Piedmont Skylark	"The sound of our jet engines becomes a song . . . the one you remember when you go back home, or the one that says you're heading to the start of something grand."	*Newsweek* November 16, 1970
Piper Aircraft Corp.	Piper Cherokee D	"You're cruising in quiet, luxurious comfort at over 140 mph."	*Sports Illustrated* April 14, 1969

F. *Aircraft (Cont'd.)*

COMPANY	PRODUCT	ADVERTISING COPY	PERIODICAL
Rolls Royce Ltd.	Rolls-Royce Engine RB.211	"One of the big points Rolls-Royce wants to prove is the quieter running of their new engine." "The fan can be slowed down on approach to landing to reduce noise. Unique three-shaft design and other sophisticated technology help make the RB.211 quieter than any other fan jet now in use."	*U.S. News & World Report* June 2, 1969
United Aircraft	Aircraft Engine JT9D	"So quiet, so smooth, that passengers don't even know there are engines on the plane."	*Forbes* February 15, 1969
		"Neighbors of airports don't have to be told what's up. It's noise. But quieter engines are coming." "And these engines will be quieter and virtually smoke-free."	*Newsweek* October 27, 1969
		"So quiet, so smooth, that passengers don't even know there are engines on the plane." "We're working to reduce fuel consumption, exhaust smoke, and noise. Our powerful new JT9D, for instance, is quieter than current jet engines."	*Newsweek* November 24, 1969

G. *Recreational Equipment*

COMPANY	PRODUCT	ADVERTISING COPY	PERIODICAL
Evinrude Division of Outboard Marine Corp.	Outboard engine	"Sound sealed quieting."	*Nation's Business* June, 1969
	Outboard engine	"The push-button new Triumph 60 is as elegant and quiet as it is efficient quick and strong."	*Popular Science* March, 1970
		"Smaller, smoother, faster, quieter, more powerful." ". . . a new Evinrude development that uses the energy of sound to get more power . . . its done quietly."	*Popular Science* October, 1970

G. Recreational Equipment (Cont'd.)

Company	Product	Advertising Copy	Periodical
		"It's so smooth and quiet when you're trolling, you may wonder whether it's even running."	*Popular Science* February, 1971
	Outboard engine: 100 hp Starflite	"No outboard ever delivered so much performance—with such quiet . . ."	*Popular Science* April, 1971
		"One of the strong points of the "100" is its internally tuned exhaust, which uses sound energy to deliver free horsepower—without making a lot of noise about it."	*National Geographic* March, 1971
Johnson Division of Outboard Marine Corp.	Outboard engine	"Silencing is improved, vibration dampened with a new isolation system."	*Look* May 13, 1969
		"Feels like more than 55 horsepower but doesn't sound like it—patented water-shield silencing encases (and cools) the exhaust housing in a sound absorbing water chamber while a unique isolation system tones out vibration."	*Look* June 10, 1967
		"You can buy other outboard motors with horsepower comparable to our new Sea-Horse 115. But you'll have to settle for something noisier . . ."	*Popular Science* April, 1970
		"If smooth, quiet, floatboating is your game, more power to you with this Sea-Horse 25."	*Look* May 5, 1970
		"The exhaust finally (and quietly) exits underwater through the prop hub."	*Popular Science* February, 1971
Kiekhaefer Mercury	Outboard engine	"Full Mercury silencing. Including Jet-Prop exhaust, sound-insulation cowl suspension, sound absorbent cowl liner, and sound deadening "wall of water" jacket surrounding the exhaust tube."	*Motor Boating* June, 1969
Outboard Marine Corporation	Outboard engine: Stern Drive	"Enjoy the economy and quiet of Jet-prop exhaust."	*Motor Boating* January, 1971

G. Recreational Equipment (Cont'd.)

COMPANY	PRODUCT	ADVERTISING COPY	PERIODICAL
	Outboard engine: Stern Drive	"Yet with all that power, you ride in a quiet comfort only possible with OMC's exclusive vibration absorbers."	*Popular Science* April, 1970
Outboard Marine Corporation	Snowmobile: Johnson Skee-Horse	"The engine is completely enclosed. So is noise, dirt and ugliness."	*Popular Science* November, 1970

Notes

PREFACE

1. Vern O. Knudson, "Noise, the Bane of Hearing," *Noise Control,* 1:3, May, 1955, p. 11.

2. M. S. Goromosov, *The Physiological Basis of Health Standards for Dwellings,* Public Health Paper #33 (Geneva: World Health Organization, 1968), p. 70.

3. G. G. Luce and Dennis McGinity, *Current Research in Sleep and Dreams,* US Public Health Service, Department of HEW, Report 1389 (Washington, D.C.: GPO, 1965), and William C. Dement, "The Effect of Partial REM Sleep and Delay Recovery," *Journal of Psychiatric Research,* 4:2, April, 1966, pp. 141-152.

4. Richard H. Mott, *et al.,* "Comparative Study of Hallucinations," *Archives of General Psychiatry,* 12:6, June, 1965, pp. 595-601, and Arnold M. Ludwig, "Auditory Studies in Schizophrenia," *American Journal of Psychiatry,* 119:8, August, 1962, pp. 122-127.

CHAPTER I. *Community Noise as a Social Problem*

1. Marshall McLuhan, *The Gutenberg Galaxy* (Toronto: University of Toronto Press, 1962), p. 83.

2. Frederick C. Kenyon, *Books and Readers in Ancient Greece and Rome* (Oxford: Clarendon Press, 1937), p. 65.

3. Zevedei Barbu, *Problems of Historical Psychology* (New York: Grove Press, 1960), pp. 19-26.

4. J. J. Raven, *The Bells of England* (London, 1906), p. 136.

5. A. E. Poole, ed., *Medieval England* (Oxford: Oxford University Press, 1958), p. 252.

6. G. T. Salusbury, *Street Life in Medieval England* (Oxford, Pen-In-Hand Publishing Co., 1948), p. 168.

7. Alexander Cohen, "Location-Design Control of Transportation Noise," *Journal of the Urban Planning and Development Division: ASCE,* 93:4, December, 1967, p. 82.

8. Walter H. Waggoner, "Jersey Calls on Airlines to Use Smokeless Jets," *New York Times,* October 10, 1970.

CHAPTER I (continued)

9. Emerson Globe, "Lip Service to Noise Control," *Architectural Record,* 134:11, November, 1964.

10. Warren Brodey, "Sound and Space," *Journal American Institute of Architects,* 42:7, July, 1964, pp. 58-60.

11. William H. Stewart, "Keynote Address," in D. W. Ward and J. Fricke, eds., *Noise as a Public Health Hazard* (Washington, D.C.: The American Speech and Hearing Association, 1969), p. 8.

12. Harland Manchester, "Rising Tide of Noise," *National Civic Review,* 53:1, Spring, 1964, pp. 418-422.

13. David Apps, "Cars, Trucks, and Tractors as Noise Sources," *Noise as a Public Health Hazard,* (Washington, D.C.: American Speech and Hearing Association, 1969), pp. 317-320.

14. Glynn Mapes, "A Vacuum's Woosh, A Car Door's Thunk Don't Just Happen," *Wall Street Journal,* September 10, 1968.

15. *The 1970 Buying Guide Issue Consumer Reports,* Consumers Union, 34:12, December, 1969, p. 57. See also *Consumer Reports,* March, 1967, for the complete vacuum cleaner report.

16. Mapes, *op. cit.*

17. Manchester, *op. cit.*

18. Howard W. Bredin, "City Noise: Designers Can Restore Quiet at a Price," *Product Engineering,* 39:24, November 18, 1968, p. 35.

19. B. G. Wald and P. H. Tope, "Household Noise Problems—Appliances," presented at the 80th Meeting Acoustical Society of America, Houston, Texas, November 3-6, 1970.

20. "The First Quiet Portable Compressor," *Sound and Vibration,* 3:5, May, 1969.

21. Armand M. Nicholi, "The Motorcycle Syndrome," *American Journal of Psychiatry,* 126:11, May, 1970, p. 1591.

22. Frederick L. Dey, "Auditory Fatigue and Predicted Permanent Hearing Defects from Rock-And-Roll Music," *The New England Journal of Medicine,* 282:9, February 26, 1970, pp. 467-470.

23. "Kiddie Cars and Tractors," *Consumer Reports* 35:11, November, 1970, pp. 632-637.

24. David C. Hodge and R. B. McCommons, "Acoustical Hazards of Children's Toys," *Journal of Acoustical Society of America,* 40:4, October, 1966; and K. Gjaevenes, "Measurements on the Impulsive Noise from Crackers and Toy Firearms," *Journal Acoustical Society of America,* 39, 1966.

25. Paul Showers, *The Listening Walk,* (New York: Thomas Y. Crowell, 1961.)

26. "Pollution Foes Picket Store over Toy that Smokes," *New York Times,* January 28, 1969.

27. The use of air-conditioning has helped lull us into a false sense of well-being concerning air pollution. Present air-conditioning filters used for homes, offices, and cars do an inadequate job of removing gaseous pollutants and particulate matter.

CHAPTER I (continued)

28. Leo Beranek, "The Effects of Noise on Man and Its Control," presented at the Atmospheric Noise Pollution and Measures for Its Control Course (Berkeley, California, June 17-21, 1968).

29. Paul Borsky, "The Effects of Noise Community Behavior," *Noise as a Public Health Hazard* (Washington, D.C.: American Speech and Hearing Association, 1969).

30. Audiometric data comparing present school age children to those of previous years suggests today's children have greater hearing loss. Evidence presented by Dr. D. M. Liscomb at National Academy of Science–National Research Council meeting of Committee on Hearing, Bioacoustics and Biomechanics Working Group 59, Environmental Noise Hazards. Washington, D.C., November 19, 1969.

31. U.S. Department of Interior, *National Resources Mission to Germany: A Special Report to the President,* October, 1966.

32. German Engineers Association (VDI), *Evaluation of Working Noise in the Neighborhood,* VDI Regulation 2058, Parts 1 and 2, August, 1968.

33. Executive Office of the President, Bureau of the Budget, Office of Management and Budget, Federal Budget, August 6, 1970 (preliminary data).

34. Clean Air Act, Title IV, Noise Pollution, Public Law 91-604, December 31, 1970.

35. Office of the White House, Memorandum for Heads of Departments and Agencies, "Aircraft Noise and Compatible Land Use in the Vicinity of Airports," March 22, 1967.

36. Veterans Administration, Information Service, Compensation and Pension Branch.

37. Department of Transportation, "Summary of Sonic Boom Claims Presented in the US to the Air Force, Fiscal Years 1957-67" (Washington, D.C., 1967). For more current information see William F. McCormack, Major, "Legal Aspects of Sonic Boom" presented at SAE/DOT Conference on Aircraft and the Environment, Washington, D.C., February 8-10, 1971.

38. Committee on the Problem of Noise, *Noise: Final Report* (London: HMSO, 1963), p. 219.

39. Stephen Ring and Larry Schein, "Practices and Utilization of Mental Health Gatekeepers in an Urban Area," presented at the 46th Meeting, American Orthopsychiatric Association, March, 1969.

40. Gladwin Hill, "Conservationists Count Their Election Victories," *New York Times,* November 5, 1970.

41. "80,000 Governments," *American Society of Planning Officials Newsletter,* 34:9, October, 1968, p. 114.

42. David Bird, "Environmental Superagency Asks City for Half Billion for Projects," *New York Times,* October 30, 1968.

43. David Bird, "Sirens Scream for Quiet's Sake," *New York Times,* December 19, 1969.

44. "City's Noise Bedlam" (editorial), *St. Louis Globe Democrat,* May 17, 1967.

CHAPTER I (continued)

45. J. Naughton, "Nixon Proposes 2 New Agencies on Environment," *New York Times*, July 10, 1970.

46. William R. Bradley, "Living with Noise Laws and Regulations," *Industrial Hygiene Foundation: Transactions Bulletin #40*, Proceedings 31st Meeting, 1966.

47. *American Airlines, Inc. et al. v. Town of Hempstead, et al.*, U.S. District Court, Eastern District of New York, 1967.

48. Joseph P. Fried, "Revised Building Code Approved by City Council," *New York Times*, October 23, 1968.

49. Committee on the Problem of Noise, *op. cit.*, pp. 36-41.

50. Department of California Highway Patrol, Vehicle Code, Sections 23130 and 27160, 1967.

51. *Ibid.*

52. J. R. Crotti, Report to the Legislature on Proposed Noise Standards for California Airports Pursuant to Public Utilities Code 21669, California Department of Aeronautics, April 1, 1970. See also "California Board Votes Noise Curb, 15 Year Program Is Adopted for Airports in State," *New York Times*, November 12, 1970.

53. Reported by Richard Metzler, "Quiet Communities for New York," presented at the annual meeting of the American Public Health Association, October 29, 1970.

54. Reported in "Noise Bureau to Conduct Studies, Submit Standards to State Control Board," *Environmental Reporter*, 1:38, January 15, 1971, p. 1000.

55. Noise Certification (H.R. 3400), Public Law 90-411, 82 Stat., 395, July 21, 1968.

56. Reported by Robert Lindsey, "FAA Acts to Cut Noise of Jetliners," *New York Times*, November 13, 1969.

57. Department of Transportation, Federal Aviation Agency Administration, Noise Standards: Aircraft Type Certification, Part 36, Chapter 1, Title 14 —Aeronautics and Space, issued January 3, 1969.

58. *Ibid.*

59. Reported by Lindsey, *op. cit.*

60. Reported by Christopher Lydon, "Higher Airport Noise Level Foreseen," *New York Times*, February 25, 1970.

61. *Ibid.*

62. "Berkeley Council Rejects Copter Patrols by Police," *New York Times*, May 31, 1970.

63. Civil Airplane Noise Reduction Retrofit Requirements Advance Notice of Propose/Rule Marking (Docket No. 10664; Notice No. 74-44), *Federal Register*, 35, November 4, 1970.

64. Robert Lindsey, "A Plan to Muffle All Jets Draws Airline Opposition," *New York Times*, February 14, 1971.

65. Reported by Laurence Marks, "Police Chiefs Back-Pedal on Noise Meters," *London Observer*, September 22, 1968.

66. *Ibid.*

CHAPTER I (continued)

67. Motor Vehicle Noise Limit, Section 386, Vehicle and Traffic Law, New York State Thruway, October 1, 1965.

68. House of Representatives, *Congressional Record,* May 2, 1966, pp. 9024-9029.

69. Stannard M. Potter, "Opening Remarks—Community Noise Control," *Noise as a Public Health Hazard* (Washington, D.C.: American Speech and Hearing Association, 1969), pp. 309-311.

70. Personal correspondence, K. L. Rose, Administrative Services Division, Memphis Police Department.

71. Ross Little, "California Laws and Regulations Relating to Motor Vehicle Noise," presented at the 78th meeting of the Acoustical Society of America, San Diego, California, November 4, 1969.

72. George J. Thiessen, "Survey of the Traffic Noise Problem," paper presented at the 69th Meeting of the Acoustical Society of America, Washington, D.C., June 2, 1965.

73. William E. Burrows, "City Backs Housing Despite Jet Fear," *New York Times,* October 26, 1967.

74. *Ibid.*

75. Reported by Edward Hudson, "Heliport Opposed by Rockefeller U.," *New York Times,* January 18, 1968.

76. Reported in "Pan Am Will Run Heliport in City," *New York Times,* October 19, 1968.

77. "86 Airports Noise Grants in 6 Months," *London Daily Telegraph,* March 10, 1968.

78. *Ibid.*

79. Elliot F. Porter, Jr., "Noise in Urban Life: The Deafening Din," *St. Louis Post-Dispatch,* October 4, 1970.

80. Reported by Deirdra Carmody, "Blast of Construction Shatters Nerves and Windows," *New York Times,* September 26, 1970.

81. Child Protection and Toy Safety Act, Public Law 91-113, 91st Congress S. 1689, November 6, 1969.

82. "Hazardous Substances: Definitions and Procedural and Interpretative Regulations," Part 191, Chapter 1, Title 21 Food and Drugs, *Federal Register,* December 19, 1970.

83. General Motors Corp., *Operation Hush: A Synopsis of the Commercial Vehicle Noise Reduction Programs by the General Motors Truck and Coach Division,* June, 1970.

84. Personnel correspondence with Mr. Ralph K. Hillquist, Noise and Vibration Laboratory, General Motors Proving Ground, Milford, Mich.

85. "200 New Garbage Trucks Will Replace Last of the Noisy Ones," *New York Times,* January 31, 1970.

86. James H. Botsford, "Control of Industrial Noise through Engineering," *Noise as a Public Health Hazard* (Washington, D.C.: American Speech and Hearing Association, 1969), p. 133.

87. James H. Botsford, "Engineering Standards and Specifications," lec-

CHAPTER I (continued)

ture presented Pennsylvania State University, College Station, Pa., July, 1967.

88. Personal discussion with Mr. Frank W. Church, Esso Research and Engineering Company.

89. House of Representatives, *Investigation and Study of Aircraft Noise Problems,* 88th Congress, 1963, H.R. 36.

90. National Aircraft Noise Symposium, Report of Proceedings. (Jamaica, N.Y., 1965).

91. Committee on Environmental Quality of the Federal Council for Science and Technology, *Noise—Sound without Value,* (Washington, D.C.: GPO, September, 1968), p. 10.

92. Reported by William E. Burrows, "Hush-Hush Agent Helps Airlines Beat Noise Ban," *New York Times,* October 20, 1967.

93. *Ibid.*

94. C. Lydon, "Higher Airport Noise Levels Foreseen," *op. cit.*

95. Reported in ""800 New Airports Needed, FAA Says," *Philadelphia Inquirer,* November 10, 1968.

96. Airport Operators Council, Inc., "An Analysis of Aircraft Noise Annoyance and Some Suggested Solutions," March, 1960, pp. 16-18.

97. Airport Operators Council, Inc., *AOCI Airport Operator's Aircraft Noise Kit,* August, 1967.

98. Committee on Environmental Quality, *op. cit.*

99. Personal discussion with Raymond Donley, physicist, noise control consultant.

100. William M. Shearer, "Acoustic Threshold Shift from Power Lawnmower Noise," *Sound and Vibration,* 2:10, October, 1968.

101. Alexander Cohen, *et al.,* "Sociocusis—Hearing Loss from Non-Occupational Noise Exposure," *Sound and Vibration,* 4:11, November, 1970, p. 18.

102. Air-Conditioning and Refrigeration Institute, "Draft of a Municipal Ordinance to Regulate Sound," 1966.

103. Warren E. Blazier, Jr., "Criteria for Control of Community Noise," *Sound and Vibration,* 2:5, May, 1968, p. 14.

104. Committee on Environmental Quality, *op. cit.,* p. 27.

105. Blazier, *op. cit.*

106. Noise data is based upon automobile road tests conducted by *Popular Science* personnel that appeared in the monthly *Popular Science* magazine between February, 1970 and January, 1971.

107. John D. Morris, "Nader Urges FTC to Ban Ads Not Backed by Scientific Tests," *New York Times,* December 12, 1970, and Richard Phalon, "Nader supports city ad proposal," *New York Times,* February 17, 1971.

108. U.S. Department of Commerce, *Snowmobiles,* (Washington, D.C.: U.S. GPO, September, 1969).

109. "Snowmobiles Sales Hit the Skids," *Business Week,* March 13, 1971.

110. R. E. Poyner and F. H. Bess, "The Effects of Snowmobile Engine Noise on Hearing," paper presented at 46th Meeting, American Speech and Hearing

CHAPTER I (continued)

Association, November 1970, and National Research Council of Canada, *Snowmobile Noise: Its Sources, Hazards and Control*, APS-477, 1970.

111. Seth S. King, "Snowbelt State Press to Regulate Snowmobiles," *New York Times*, March 2, 1971.

112. Walsh-Healey Public Contracts Act, Part 50-204, Safety and Health Standards for Federal Supply Contracts, Section 50-204.10 Occupational Noise Exposure, *Federal Register*, March 20, 1969.

113. Botsford, "Control of Industrial Noise through Engineering," *op. cit.*

114. Federal Aviation Act of 1958 (S.R. No. 1811, 85th Congress, 2nd Session, 1958).

115. Nicholas E. Golovin, "The Public and National Noise Standards," presented at the 76th Meeting of the Acoustical Society of America, Cleveland, O., November 21, 1968.

116. Quoted in Evert Clark, "Noise Called Bar to New Airports," *New York Times*, October 5, 1967.

117. Cecil M. Mackey, speech to the Fourth Annual Meeting of the American Institute of Aeronautics and Astronauts, Anaheim, California, reported *New York Times*, October 25, 1967.

118. *Ibid.*

119. U.S. Council on Environmental Quality, *Environmental Quality*, First Annual Report (Washington, D.C.: GPO, 1970).

120. Reported in "An Opening Attack on the Decibel Din: New Law Seeks to Control Aircraft Noise," *Conservation Foundation Newsletters*, August 30, 1968, pp. 4-5.

121. Quoted in *ibid.*

122. Quoted in *ibid.*

123. House of Representatives Committee on Interstate and Foreign Commerce, *Aircraft Noise Abatement Hearing* (Washington, D.C.: GPO, 1968), statement by Hon. Byron G. Rogers, Representative, Colorado.

124. Noise Certification, *op. cit.*

125. Reported by Christopher Lydon, "Funds for the SST Is Voted in House by a Slim Margin," *New York Times*, May 28, 1970. Senate Defeats SST Funding (52 Nays to 41 Yeas) *Congressional Record*, 116:192, December 3, 1970.

126. "Senate SST Filibuster Pledged Confrees Will Convene Today," *New York Times*, December 10, 1970.

127. House Approves Extending SST Appropriation to March 31, 1971 (180 yeas to 37 nays), *Congressional Record*, 116:211, December 31, 1970, M12605-6, and Senate Approves Extending SST Appropriating to March 31, 1971 (Voice vote), *Congressional Record*, 116:212, January 2, 1971, S21779.

128. The U.S. House of Representatives by a vote of 215 yeas to 204 nays agreed to an amendment removing further funding of the SST.(*Congressional Record*, 117:38, March 18, 1971, pp. 1748-1749). The U.S. Senate by a vote of 51 nays to 46 yeas rejected an amendment restoring funds for continuing construction of the SST prototype (*Congressional Record*, 117:42, March 24, 1971, pp. 3868-3869).

CHAPTER I (continued)

129. U.S. White House, in "Text of Statement by President Nixon on SST," *New York Times,* December 6, 1970.

130. Reported in "Project Chief Lobbies Hard to Sell the SST," *New York Times,* July 20, 1970.

131. Reported by F. Zimmerman, "Supersonic Snow Job," *Wall Street Journal,* February 9, 1967.

132. Victor Block, "The Supersonic Transport and You," *Science Digest,* 60:1, July, 1966.

133. Karl D. Kryter, "Sonic Boom—Results of Laboratory and Field Studies," *Noise as a Public Health Hazard* (Washington, D.C.: American Speech and Hearing Association, 1969).

134. B. Lundberg, *op. cit.,* and B. Lundberg, "The Acceptable Nominal Sonic Boom Overpressure in SST Operation," *Noise as a Public Health Hazard* (Washington, D.C.: American Speech and Hearing Association, 1969), pp. 278-297.

135. *Report of the Secretary of the Interior of the Special Study Group on Noise and Sonic Boom in to Man, op. cit.*

136. Reported in Lydon, *op. cit.*

137. Reported by Christopher Lydon, "Senate Approves Two Curbs on SST," *New York Times,* December 3, 1970.

138. David Bird, "Rickles Asks Noise Limits to Ban Jets from City," *New York Times,* December 30, 1970.

139. Clyde H. Farnsworth, "Conference on Sonic Boom Told Noise Can't Be Designed Away," *New York Times,* February 4, 1970.

140. Lockheed Aircraft Corporation, Advertisement, *Newsweek,* October 24, 1966.

141. *Ibid.*

142. William H. Stewart, "Keynote Address," *Noise as a Public Health Hazard,* (Washington, D.C., American Speech and Hearing Association, 1969), p. 10.

143. Lockheed, *op. cit.*

144. Reported by Albert H. Odell, "Jet Noise at John F. Kennedy International Airport," presented before the Panel on Jet Aircraft Noise, Washington, D.C., October 29, 1965.

145. Chicago, City of Chicago Noise Ordinance, July 1, 1971.

146. Reported by David Bird, *op. cit.*

147. "Incompatible Land Use—Rosedale Long Island, N.Y.—J. F. K. Airport," *Newsletter, National Aircraft Noise Abatement Council,* 9:3, March 15, 1966, pp. 6-7.

148. *Ibid.*

149. Lewis S. Goodfriend, "Control of Noise through Propaganda and Education," *Noise as a Public Health Hazard* (Washington, D.C.: American Speech and Hearing Association, 1969), p. 342.

150. William A. Shurcliff, *Newsletter #9,* Citizen's League Against The Sonic Boom, October 10, 1967.

CHAPTER I (continued)

151. William A. Shurcliff, *S/S/T and Sonic Boom Handbook* (New York: Ballantine Books, 1970).

152. Alex Baron, "The Noise Receiver: The Citizen," *Sound and Vibration,* 2:5, May, 1968, pp. 8-11.

153. Martin Gansberg, "Grants Will Aid Pollution Study," *New York Times,* July 6, 1970.

154. "Environmental Health Can Expand under New Regime," *Environmental Health Letter,* 8:24, December 15, 1969, p. 1.

155. Senator Mark Hatfield, "Compilation of State and Local Ordinances on Noise Control," *Congressional Record,* 115:176, October 29, 1969, pp. E9031-E9112.

156. "Ssh!" *Newsweek,* 74:11, September 8, 1969.

157. "How Your Congressman Voted on Critical Environmental Issues," prepared by the League of Conservation Voters, Brochure 1970 (undated).

158. Child Protection and Toy Safety Act, *op. cit.*

159. Reported in John D. Morris, "Consumer Groups Seek Pre-Holiday Ban on 8 Toys," *New York Times,* November 19, 1970.

160. "Bang, bang, you're deaf!" *Consumer Reports,* 35:11, November, 1950.

161. David C. Hodge, *op. cit.*

162. "Hazards Substances: Definitions and Procedural and Interpretative Regulations," *op. cit.*

163. Goodfriend, *op. cit.*

164. *Sound* replaced *Noise Control,* also sponsored by the Acoustical Society of America. Lasting two years, the final *Sound* issued appeared November/December, 1963.

165. "Noise and Vibration Control Materiels," *Sound and Vibration,* 4:7, July, 1970.

166. "Noise and Vibration Control Systems," *Sound and Vibration,* 4:8, August, 1970.

167. "Hearing Protective Devices," *Sound and Vibration,* 4:11, November, 1970.

168. "Dynamic Measurement Instrumentation Buyer's Guide," *Sound and Vibration,* 4:12, December, 1970.

169. Circulation Policy of *Sound and Vibration,* found inside front cover of each issue.

170. Jan P. Norbye and Jim Dunne, "The '70 Personal Cars Combine Luxury and Sportiness," *Popular Science,* 196:2, February, 1970.

171. Al Lees, "New Ideas for Noise Control at Home," *Popular Science,* 196:9, September, 1970.

172. "Table Saws," *Consumer Reports,* 35:5, May, 1970.

173. "The Little Cars," *Consumer Reports,* 36:1, January, 1971, pp. 8-17.

174. See Jan P. Norbye and Jim Dunne, "From Japan: Little 2-Door Sedans with a Low, Low Price," *Popular Science,* 197:6, December, 1970, p. 36, and Norbye and Dunne, "The Pinto and Vega: 10,000 Miles Later," *Popular Science,* 198:1, January, 1971, p. 14.

CHAPTER I (continued)

175. Robert A. Baron, *The Tyranny of Noise* (New York: St. Martin's Press, 1970).

176. Theodore Berland, *The Fight for Quiet* (Englewood Cliffs, N.J.: Prentice-Hall, 1970).

177. Henry Still, *In Quest of Quiet* (Harrisburg, Pa.: Stackpole Books, 1970).

CHAPTER II. *What Is Noise?*

1. Committee on the Problem of Noise, *op. cit.,* p. 2.

2. Joseph Sataloff, *Industrial Deafness: Hearing, Testing and Noise Measurement* (New York: McGraw-Hill, 1957).

3. Alexander Cohen, "Effects of Noise on Man," *Journal of the Boston Society of Civil Engineers,* 52:1, January, 1965, p. 75.

4. Arnold Peterson and Ervin Gross, eds., "What Are Noise and Vibration?" *Handbook of Noise Control* (West Concord: General Radio Company 1967), pp. 3-75.

5. Allan Bell, *Noise: An Occupational Hazard and Public Nuisance,* Public Health Paper #30 (Geneva: U.N. World Health Organization, 1966), p. 10.

6. Donald C. Gasaway, "Aeromedical Significance in Noise Exposures Associated With the Operation of Fixed- and Rotary-Winged Aircraft" (Brooks Air Force Base, Texas, USAF School of Aerospace Medicine, November 1965). (Mimeographed)

7. Peterson and Gross, eds., *Handbook of Noise Control,* Based on American National Standards Institute, Sound Level Meter Standard, S1.4-1961.

8. Second Intersociety Committee on Guidelines for Noise Exposure Control, "Guidelines for Noise Exposure," January 23, 1970, p. 11 (mimeographed).

9. *Federal Register,* 34:96, May 10, 1969, pp. 7946-7949, and *Federal Register,* 36:105, May 29, 1971.

10. Karl D. Kryter, *et al.,* "Hazardous Exposure to Intermittent and Steady-State Noise," *Journal Acoustical Society of America,* 39:3, March, 1966, pp. 451-464.

11. Joseph Sataloff, Lawrence Vassallo, Hyman Menduke, "Hearing Loss from Exposure to Interrupted Noise," *Archives of Environmental Health,* 18:6, June, 1969, pp. 972-981.

12. James H. Botsford, "Scales for Expressing Noise Level-Damage Risk," presented at the Symposium: Evaluating the Noises of Transportation, University of Washington, Seattle, Wash., March 26–28, 1969, p. 6.

13. *Ibid.*

14. Karl D. Kryter, "The Meaning and Measurement of Perceived Noise Level," *Noise Control,* 6:5, September–October, 1960, pp. 12-27.

15. A. H. Odell, *op. cit.*

16. Committee on the Problem of Noise, *op. cit.,* pp. 207-209.

17. Leo Beranek and R. Wayne Rudmose, "Sound Control in Airplanes," *Journal Acoustical Society of America,* 19:2, March, 1947, pp. 357-363.

18. F. J. Langdon and W. E. Scoles, *The Traffic Noise Index: A Method of*

CHAPTER II (continued)

Controlling Noise Nuisance, Building Research Station Current Papers 38/68, April, 1968, pp. 2-3.

19. I. D. Griffiths and F. J. Langdon, "Subjective Response to Road Traffic Noise," Building Research Current Papers, 37/68, April, 1968, p. 16.

20. James H. Botsford, "Using Sound Levels to Gauge Human Response to Noise," *Sound and Vibration,* 3:10, October, 1969, pp. 23-25.

CHAPTER III. *The Nuisances and Hazards of Noise*

1. M. L. Heideman, Jr.'s quantitative recasting of the quantal (all-or-none) definition given by WHO, in *The First Ten Years of the World Health Organization* (Geneva: World Health Organization, 1958).

2. Committee on the Problem of Noise, *op. cit.,* p. 8.

3. In part adapted from an article by Frank Stead, "Levels in Environmental Health," *American Journal of Public Health,* 50:3, March, 1960, pp. 312-315.

4. "Helicopter Noise Blamed in Part for 2 Deaths at Kennedy Train," *New York Times,* October 30, 1968.

5. "Noise Conference in Massachusetts," *National Aircraft Noise Abatement Council Newsletter,* 19:7, July 15, 1968, p. 5.

6. Paul S. Veneklasen, "Community Noise Control," *Noise as a Public Health Hazard* (Washington, D.C.: American Speech and Hearing Association, 1969), p. 361.

7. "Driver Accident Rate—Hearing vs. Deaf," *New York Times,* March 19, 1968, and "Deaf People Called Excellent Drivers," *New York Times,* February 18, 1970, and F. Burg *et al.,* "Licensing the Deaf Driver," Archives of Environmental Health, 91:3, March, 1970.

8. Personal correspondence with Mr. Robert Camp U.S. Army Aeromedical Research Laboratory, Fort Rucker, Alabama.

9. J. M. Pickett, "Message Constraints, A Neglected Factor in Predicting Industrial Speech Communication," *Noise as a Public Health Hazard* (Washington, D.C.: American Speech and Hearing Association, 1969), pp. 121-128. See also M. I. Freed, "Our Deaf Employees Are Not Handicapped," *Rehabilitation Record* 3:3, May–June, 1962, p. 35.

10. "Sonic Boom Damage in the National Park System," Report to the Secretary, Department of the Interior, January 10, 1967, p. 2. (mimeographed)

11. J. L. Hess, "French Investigate Deaths of Three Linked to Superjets' Boom," *New York Times,* August 3, 1967.

12. *Ibid.*

13. Hallowell Davis and Richard S. Silverman, *Hearing and Deafness* (rev. ed.; New York: Holt, Rinehart and Winston, 1962), p. 511.

14. *Ibid.*

15. McCay Vernon, "Sociological and Psychological Factors Associated with Hearing Loss," *Journal of Speech and Hearing Research,* 12:3, September, 1969, pp. 541-563.

16. H. H. Kronenberg and G. D. Blake, "Young Deaf Adults: An Occupa-

CHAPTER III (continued)

tional Survey" (Washington, D.C.: Vocational Rehabilitation Administration Department of HEW, 1966); and *ibid.*

17. Leo Beranek, "Noise," *Scientific American,* 215:6, December, 1966, p. 68.

18. Paul Michael, "Noise in the News," *American Industrial Hygiene Association,* 26:6, November–December, 1965, pp. 615-168. See also Bruce and A. Welch, eds., *Physiological Effects of Noise,* (New York: Plenum Press, 1970).

19. *Noise in Hospitals: An Acoustical Study of Noises Affecting The Patient* (Washington, D.C.: U.S. Public Health Service, 1963).

20. Donald Broadbent, "Effects of Noise on Behavior," *Handbook of Noise Control,* Cyril Harris, ed. (New York: McGraw-Hill, 1957), pp. 10-8 to 10-9.

21. K. F. H. Murrell, *Ergonomics* (London: Chapman and Hall, 1965), p. 286.

22. Gerd Jansen, "Effects of Noise on Physiological State," *Noise as a Public Health Hazard* (Washington, D.C.: American Speech and Hearing Association, 1969), pp. 89-99.

23. A. Carpenter, "Effects of Noise on Performance and Productivity," *The Control of Noise* (London: HMSO, 1962), p. 30.

24. *Noise Abatement: A Public Health Problem* (Luxembourg: Council of Europe, 1965), p. 9.

25. N. F. Svadlrouskaya, "Effect of Long-Term Noise on Cerebral Oxidation Processes in Albino Rats" (Russian), *Gigiena I. Sanitaria,* Volume 7, 1967.

26. "Public Health Aspects of Housing in the USSR," *WHO Chronicle,* 20:10, October, 1966, p. 357.

27. Murrell, *op. cit.,* p. 292.

28. Gunther Lehmann, "Noise and Health," *The UNESCO Courier,* July, 1967, p. 31.

29. *Ibid.*

30. Jansen, *op. cit.,* pp. 94-96.

31. *Ibid.*

32. *Ibid.,* pp. 96-98.

33. Broadbent, *op. cit.*

34. Karl Kryter, "Effects of Noise on Man," *Journal of Speech and Hearing Disorders* (Monograph, Supplement No. 1), 1950. See also Kary Kryter, *The Effects of Noise on Man* (New York: Academic Press, 1970).

35. Jansen, *op. cit.,* pp. 89-93.

36. *Ibid.,* p. 96. See also Gerd Jansen, "Relationship between Temporary Threshold Shift and Peripheral Circulatory Effects of Sound," *Physiological Effects of Noise,* B. and A. Welch, eds. (New York: Plenum Press, 1970).

37. Aram Glorig, *Noise and Your Ear* (New York: Grune & Stratton, 1958).

38. C. Bragdon, "Noise: A Syndrome of Modern Society," *Scientist and Citizen,* 10:3, March, 1968, p. 32.

39. Fred W. Braun, "Hearing Conservation in Industry," *Noise Control,* 4:7, July, 1958, p. 37.

40. Department of Health, Education & Welfare, "Characteristics of Persons with Impaired Hearing" (Washington, D.C., 1967), PHS publication #1000, Series 10, Number 35.

CHAPTER III (continued)

41. Quoted in House of Representatives, *Congressional Record,* October 19, 1966, p. 26777.

42. Aram Glorig, "Industrial Noise and The Worker," *Noise as a Public Health Hazard* (Washington, D.C.: American Speech and Hearing Association, 1969) pp. 105-109.

43. Committee on Environmental Quality of the Federal Council for Science and Technology, *Noise—Sound without Value,* September, 1968, p. 32.

44. J. D. Dougherty and O. Welsh, "Community Noise and Hearing Loss," *New England Journal of Medicine,* 127:14, October 6, 1966, p. 759.

45. Murrell, *op. cit.,* p. 286.

46. Committee on the Problem of Noise, *op. cit.,* p. 125.

47. Department of the Army Technical Bulletin (TB MED 251), *Noise and Conservation of Hearing,* January, 1965.

48. Air Force Regulation (AFR 160-3A) *Hazardous Noise Exposure,* June 27, 1960.

49. American Academy of Ophthalomology and Otolaryngology, Subcommittee on Noise in Industry, *Guide for Conservation of Hearing in Noise* (Los Angeles, 1957). This guide has subsequently been revised in 1967.

50. Walsh-Healey Public Contracts Act, *ibid.*

51. N. Rosenwinkel and K. Stewart, "Relationship of Hearing Loss to Steady State Noise Exposure," *American Industrial Hygiene Quarterly,* 18, 1957.

52. W. Dixon Ward, "The Effects of Noise on Hearing Thresholds," *Noise as a Public Health Hazard* (Washington, D.C.: American Speech and Hearing Association, 1969), pp. 40-48.

53. William L. Baughn, "Noise Control—Percent of Population Protected," *International Audiology,* 5:9, September, 1966, pp. 331-338.

54. Alexander Cohen *et al.,* "Sociocusis—Hearing Loss from Non-Occupational Noise Exposure," *Sound and Vibration* 4:11, November, 1970.

55. Bell, *ibid.,* p. 30, and I. I. Ponomarenko, "Standards of High-Frequency Industrial Noise for Young Workers," *Russian Industrial and Hygiene Sanitation* (Translation), 1969, pp. 192-196.

56. Aram Glorig, "Effects of Noise on Man," *Journal American Medical Association,* 196:10, June 6, 1966, p. 339.

57. Ward, *ibid.,* p. 46.

58. Joseph Sataloff and John A. Zapp, "The Environment in Relation to Otologic Disease," *Archives of Environmental Health,* 10:3, March, 1965, pp. 403-415.

59. Allan Bell, *Noise: An Occupational Hazard and Public Nuisance,* Public Health Paper #30 (Geneva: U.N. World Health Organization, 1966), p. 24.

60. Joseph Sataloff, "Temporary and Permanent Hearing Loss," *Archives of Environmental Health,* 10:1, January, 1965, p. 410.

61. W. D. Ward and D. Nelson, "TTS and PTS in Chinchillas," *Journal Acoustical Society of America.* Presented at Acoustical Society of America, Atlantic City, N.J., April 22, 1970.

62. Karl Kryter, "The Effects of Noise on Man," *Noise as a Public Health*

CHAPTER III (continued)

Hazard (Washington, D.C.: American Speech and Hearing Association, 1969), p. 38.

63. Bell, *op. cit.*

64. N. Barnett and B. Erickson, "The Sonic Environment and Its Effects on Man," *School Environmental Research—2* (Ann Arbor: University of Michigan, 1965).

65. Harvey Fletcher, *Speech and Hearing in Communication* (New York: Van Nostrand Co., 1953).

66. Beranek, *op. cit.*, p. 70.

67. John C. Webster, "SIL—Past, Present, and Future," *Sound and Vibration,* 3:8, August, 1969, pp. 22-26.

68. W. Niemeyer, "Speech Discrimination in Noise Induced Deafness," *International Audiology,* 6:6, June, 1967, p. 46.

69. Alexander Cohen, "Effects of Noise on Man," *Journal of the Boston Society of Civil Engineers,* 52:1, January, 1965, pp. 83-84.

70. John C. Webster, "The SIL—Past, Present, and Future," *Sound and Vibration,* 3:8, August, 1969.

71. John C. Webster and M. Lepor, "Noise—You Can Get Used to It," *Journal Acoustical Society of America,* 45:3, March, 1969, pp. 751-757.

72. J. Ferguson and W. Dement, "The Effect of Variations in Total Sleep Time on the Occurrence of Rapid Eye Movement Sleep in Cats," *Electroencephology and Clinical Neurophysiology,* 22, July, 1967, pp. 2-10.

73. W. Dement, "Recent Studies on the Biological Role of Rapid Eye Movement Sleep," *The American Journal of Psychiatry,* 22:10, October, 1965, pp. 404-408.

74. Ferguson and Dement, *op. cit.*

75. Dement, *op. cit.*

76. See G. W. Vogel, "REM Deprivation," *Archives of General Psychiatry,* 18:3, March, 1968, pp. 287-329, and E. Hartmann, "The 90 Minute Sleep Dream Cycle," *Archives of General Psychiatry,* 18:3, March, 1968, pp. 280-286.

77. Luce and McGinity, *ibid.*

78. Samuel Rosen, "Noise Pollution: A Need for Action," *Medical Tribune,* January 4, 1968.

79. George Thiessen, "Community Noise—Surface Transportation," *Sound and Vibration,* 2:4, April, 1968, pp. 14-16.

80. Jansen, *op. cit.*, p. 94.

81. Jerome S. Lukas and Karl Kryter, *A Preliminary Study of the Awakening and Startle Effects of Simulated Sonic Booms,* Stanford Research Institute, April, 1968, pp. 35-38.

82. Reported in E. Grandjean, *Fitting the Task to the Man: An Ergonomic Approach* (London: Taylor & Francis Ltd., 1969).

83. George Thiessen, "Effects of Noise during Sleep," *Physiological Effects of Noise,* B. Welch, ed. (New York: Plenum Press, 1970), p. 271.

84. Seminaire interregional sur l'habitat dans ses rapports avec la sante public, World Health Organization, PA/185.65. See summary in *WHO Chronicle,* October, 1966.

CHAPTER III (continued)

85. Committee on the Problem of Noise, *op. cit.*, p. 32. See also Harold L. Williams, "Auditory Stimulation, Sleep Loss and the EEG Stages of Sleep," B. and A. Welch, eds., *Physiological Effects of Noise* (New York: Plenum Press, 1970), p. 278.

86. Karl Kryter, *op. cit.*

87. David Glass *et al.*, "The Psychic Cost of Adaptation to an Environmental Stressor," November, 1968 (pre-publication draft).

88. Michael Rodda, *Noise in Society* (London: Oliver & Boyd, 1967), pp. 52-58.

89. Carpenter, *op. cit.*, p. 306.

90. Donald Broadbent, "Effects of Noises of High and Low Frequency on Behavior," *Ergonomics*, 1:1, November, 1967, pp. 21-29.

91. H. C. Weston and S. Adams, Industrial Health Research Board Report, No. 70 (London, 1935).

92. D. Eldredge, *et al.*, "Proactive Inhibition, Recency and Limited-Channel Capacity under Acoustic Stress," *Perceptual Motor Skills*, August, 1967.

93. D. Lehman, *et al.*, "An Investigation of the Effects of Various Noise Levels as Measured by Psychological Performance and Energy Expenditure," *Journal of School Health*, 35:3, March, 1965.

94. J. E. Hoffman, "The Effect of Noise on Intellectual Performance as Related to Personality and Social Factors in Upper Division High School Students," *Dissertation Abstracts*, 27:1658, No. 6-A, 1966.

95. John F. Corso, *The Effects of Noise on Human Behavior* (Wright Air Development Center: WADC Technical Report, 53-81, 1952).

96. Barbara R. Slater, "Effects of Noise on Pupil Performance, *Journal of Educational Psychology*, 59:4, August, 1968, pp. 239-243.

97. Broadbent, *op. cit.*, pp. 10-11.

98. Committee on the Problem of Noise, *op. cit.*, p. 12.

99. Quoted in G. Lehman, *op. cit.*, p. 31.

100. *Ibid.*, pp. 30-31.

101. John M. Mecklin, "Its' Time to Turn Down All That Noise," *Fortune*, October, 1969.

102. Michael Rodda, *Noise in Society* (London: Oliver & Boyd, 1967), p. 60.

103. *Noise Abatement: A Public Health Problem*, *op. cit.*, p. 13.

104. H. Warren Dunham, "Epidemiology of Psychiatric Disorders as a Contribution to Medical Ecology," *Archives of General Psychiatry*, 14:1, January, 1966.

105. G. Jansen, *op. cit.*, p. 93.

106. Hallowell Davis, ed., *Auditory and Non-Auditory Effects of High Intensity Noise*, U.S. Naval School of Aviation Medicine, June, 1958.

107. Alexander Cohen, "Noise and the Psychological State," *Noise as a Public Health Hazard* (Washington, D.C.: American Speech and Hearing Association, 1969), pp. 74-88.

108. E. E. Lieber, *Occupational Health*, (London: Business Publications, 1964).

109. C. Michalova and V. Hrubes, "Frequency of Neuroses and Psycho-

CHAPTER III (continued)

somatic Diseases Mainly Hypertension under the Influence of Noise" (unpublished report).

110. Paul Borsky, "The Effects of Noise on Community Behavior," *Noise as a Public Health Hazard* (Washington, D.C.: American Speech and Hearing Association, 1969), p. 192.

111. George Bugliarello, *Noise Pollution: A Review of Its Techno-Sociological and Health Aspects,* Biotechnology Program, Carnegie-Mellon University, February 1, 1968.

112. Leo Beranek, "Noise," *op. cit.,* p. 68.

113. E. Grandjean, "Effects of Noise on Man," *Noise as a Public Health Hazard* (Washington, D.C.: American Speech and Hearing Association, 1969), pp. 99-104.

114. Robert Ardrey, *The Territorial Imperative* (New York: Atheneum Press, 1966).

115. Edward T. Hall, *The Hidden Dimension* (New York: Doubleday, 1966).

116. Edward T. Hall, "Proxemics: The Study of Man's Spatial Relations," *Man's Image in Medicine and Anthropology* (New York International University Press, 1963), pp. 423-443.

117. W. Goldfarb and J. Mintz, "The Schizophrenic Child's Reaction to Time and Space," *Archives of General Psychiatry,* 5, 1961, pp. 535-543.

118. M. J. Horowitz, "Spatial Behavior and Psychopathology," *The Journal of Nervous and Mental Disease,* 140:1, January, 1968, pp. 24-35.

119. "Real Estate Here Is Still Great Bargain Traveler Finds," *Philadelphia Inquirer,* July 14, 1968.

120. A. D. Colman, "Territoriality in Man: A Comparison of Behavior in Home and Hospital," *American Journal of Orthopsychiatry,* 38:4, April, 1968.

121. J. Kaufman, "Control through Laws and Regulations," *Noise as a Public Health Hazard* (Washington, D.C.: American Speech and Hearing Association, 1969), pp. 327-341.

122. Robert Sherrill, "The Jet Noise Is Getting Awful," *New York Times,* January 14, 1968.

123. Charles Harr, "Airport Noise and the Urban Dweller: A Proposed Solution" (New York: Speech, Practicing Law Institute, May 10, 1968), p. 3.

124. *Ibid.*

125. "Families Ask $400,000 for Jet Noise" (S/V News), *Sound and Vibration,* 1:12, December, 1967.

126. H. K. Smith, "Jet Planes Hurt Hearing Bring Protest to Capitol," *Philadelphia Bulletin,* March 26, 1967.

127. "Boston Sues 19 Airlines over Aircraft Noise (S/V News), *Sound and Vibration,* 4:10, October, 1970.

128. John L. Hesse, "French Towns Sue over Airport Noise," *New York Times,* July, 1970.

129. "Summary of Sonic Boom Claims Presented in the U.S. to the Air Force, Fiscal Years 1956–67" (Washington D.C.: Department of Transportation, 1967).

Bibliography

Anon. "Air Pollution by Noise," *Lancet,* 1, May 2, 1970, pp. 928-929.

Anon., "Public Health Aspects of Housing in the U.S.S.R.," *World Health Organization Chronicle,* 20:10, October, 1966.

Anon. "Ssh!" *Newsweek,* 74:11, September 8, 1969.

Anticaglia, Joseph R. "Noise In Our Overpolluted Environment," *Physiological Effects of Noise,* New York: Plenum Press, 1970.

Bailey, A. "Noise Is a Slow Agent of Death," *New York Times Magazine,* November 23, 1969.

Barnett, N., and B. Erickson, "The Sonic Environment and Its Effects on Man," *School Environmental Research—2.* Ann Arbor: University of Michigan, 1965.

Baron, Robert A. *The Tyranny of Noise.* New York: St. Martin's Press, 1970.

Bell, Allan. *Noise: An Occupational Hazard and Public Nuisance,* Public Health Paper #30. Geneva: U.N. World Health Organization, 1966.

Beranek, Leo. "Noise," *Scientific American,* 215:6, December, 1966.

Berland, Theodore. *The Fight for Quiet.* Englewood Cliffs, N.J.: Prentice-Hall, 1970.

Bragdon, C. "Noise: A Syndrome of Modern Society," *Scientist and Citizen,* 10:3, March, 1968.

———. *Noise Pollution: The Unquiet Crisis.* Philadelphia: University of Pennsylvania Press, 1971.

Bugliarello, George. *Noise Pollution: A Review of Its Techno-Sociological and Health Aspects.* Biotechnology Program, Carnegie-Mellon University, February 1, 1968.

Burns, William. *Noise and Man.* London: William Clowes and Sons Limited, 1968.

Chapman, J. S. "The Sound of Noise, and the Fury," *Archives of Environmental Health* (Chicago), 20, May, 1970, pp. 612–613.

Cohen, Alexander. "Effects of Noise on Man," *Journal of the Boston Society of Civil Engineers,* 52:1, January, 1965.

Committee on Environmental Quality, Federal Council for Science and

Technology. *Noise—Sound without Value.* Washington, D.C.: GPO, September, 1968.

Council on Environmental Quality. *Environmental Quality,* The First Annual Report of the Council on Environmental Quality. Washington, D.C.: GPO, August, 1970.

Dickerson, David O., ed. *Transportation Noise Pollution: Control and Abatement.* Prepared under contract with National Aeronautic and Space Administration, 1970.

Farr, Lee E. "Medical consequences of environmental home noises," *Journal American Medical Association,* 202:3, October 16, 1967.

Goldsmith, John R., and Erland Jonsson. "Effects of Noise on Health in the Residential and Urban Environment." Prepared for the American Public Health Association, August, 1969. Mimeographed.

Grandjean, E. "Effects of Noise on Man," *Noise as a Public Health Hazard.* Washington, D.C.: American Speech and Hearing Association, 1969.

———. *Fitting the Task to the Man.* London: Taylor & Francis Ltd., 1969.

Kryter, K. D. "The Effects of Noise on Man." Monograph Supplement No. 1, *The Journal of Speech and Hearing Disorders.* American Speech and Hearing Association, 1950.

———. "The Effects of Noise on Man," *Noise as a Public Health Hazard.* Washington, D.C.: American Speech and Hearing Association, 1969.

———. *The Effects of Noise on Man.* New York: Academic Press, 1970.

Lehmann, Gunther. "Noise and Health," *The UNESCO Courier,* July, 1967.

Lieber, E. *Occupational Health.* London: Business Publications, 1964.

Manchester, Harland. "Rising Tide of Noise," *National Civic Review,* 53:1, Spring, 1964.

Maryland, Committee for the Relocation of Prince George County Airport. Noise and Safety Study. February, 1969. Mimeographed.

Michael, Paul. "Noise in the News," *American Industrial Hygiene Association,* 26:6, November–December, 1965.

Murrell, K. F. H. *Ergonomics.* London: Chapman and Hall, 1965.

National Academy of Science–National Research Council, *Report on Physical Effects of the Sonic Boom,* February, 1968.

Robinson, D. W. *An Outline Guide to Criteria for the Limitation of Urban Noise,* National Physical Laboratory, Aerodynamics Division (England), NPL Aero Report Ac 39, March, 1969.

Rodda, Michael. *Noise in Society.* London: Oliver & Boyd, 1967.

SAE/DOT Conference on Aircraft and the Environment, *Proceedings,* Society of Automotive Engineers, February 8–10, 1971.

Shurcliff, William A. *S/S/T and Sonic Boom Handbook.* New York: Ballantine Books, 1970.

Stewart, William H. "Keynote Address," *Noise as a Public Health Hazard.* Washington, D.C.: American Speech and Hearing Association, 1969.

Still, Henry. *In Quest of Quiet.* Harrisburg, Pa.: Stackpole Books, 1970.

Stromentov, Constantin. "The Architects of Silence," *UNESCO Courier,* 20:7, July, 1967.

U.S. Department of Commerce, Commerce Technology Advisory Board.

The Noise Around Us: Including Technical Backup. COM-71-00147. September, 1970.

Ward, Dixon W., and James Fricke, eds. *Noise as a Public Health Hazard.* Washington, D.C.: The American Speech and Hearing Association, ASHA Report No. 4, February, 1969.

Wilson Committee on the Problem of Noise. *Noise: Final Report.* London: HMSO, 1963.

Yerges, Lyle F. *Sound, Noise and Vibration.* New York: Van Nostrand, 1969.

Bauer, R. A. "Predicting the Future," in *Transportation Noises,* ed., J. D. Chalupnik. Seattle: University of Washington Press, 1970. pp. 245–256.   MEASUREMENT

Beranek, Leo. *Acoustics.* New York: McGraw-Hill, 1954.

——. *Acoustic Measurements.* New York: John Wiley, 1949.

Bishop, D. E., and M. A. Simpson. *Noise Exposure Forecast for 1967, 1970 and 1975 Operations at Selected Airports.* Bolt, Beranek, Newman, Inc., September, 1970. Prepared for the Federal Aviation Agency.

Bolt, Beranek, and Newman, Inc. *Procedure for Developing Noise Exposure Forecast Areas for Aircraft Flight Operations.* Prepared under contract for the Federal Aviation Agency, Department of Transportation, August, 1967.

Broch, J. T. *Acoustics Noise Measurements.* Bruel and Kjaer, Denmark, February, 1969.

Donley, R. "Measurement of Community Noise" (draft proposal of American National Standards Institute, Committee S3-W50). Presented Acoustical Society of America, San Diego, California, November 4, 1969.

Ford, R. D. *Introduction to Acoustics.* New York: Elsevier Publishing Co., 1970.

Harris, C. J. "Absorption of Sound in Air and Humidity and Temperature," *Journal of the Acoustical Society of America,* 40, 1966, p. 148.

Hewlett-Packard Co. *Acoustics Handbook.* Palo Alto, Calif., 1968.

Hillquist, Ralph K. "Test-Site Measurement of Noise Emitted by Engine-Powered Equipment" (Draft proposal of American National Standards Institute, Committee S3-W50). Presented Acoustical Society of America, San Diego, California, November 4, 1969.

Kobrynski, M. "General Method for Calculating the Sound Pressure Field Emitted by Stationary or Moving Jets." Presented at the Symposium on Aerodynamic Noise, Toronto, May 20–21, 1968. (Chastillon, France: Office National d'Etudes et de Recherches Aerospatiales, 1968.) T.P. No. 578.

Koppe, E. W., K. R. Matschat, and E. A. Muller. "Abstract of a Procedure for the Description and Assessment of Aircraft Noise in the Vicinity of an Airport," *Acustica,* 16, 1966, pp. 251–253.

Kryter, K. D. *Review of Research and Methods of Measuring the Loudness and Noisiness of Complex Sounds.* NASA, April, 1966. Report No. N66-21098.

McKaig, M. B. "Graphical Determination of Community-Noise Contour Coordinates," The Boeing Co., Renton, Washington, Report No. D6-9582TN, June 28, 1963. N68-32205. Mimeographed.

Mason, W. P., and R. N. Thurston. *Physical Acoustics: Principles and Methods.* New York: Academic Press, 1970.

Parkin, P. H., and W. E. Scoles. "Air-to-Ground Sound Propogation," *Journal of the Acoustical Society of America,* 26:6, 1954.

Parkin, P. H. "Propogation of Sound in Air," *The Control of Noise.* London: HMSO, 1962.

Partridge, Gordon R. "A Noise Exposure Meter." Presented at the 80th Meeting Acoustical Society of America, Houston, Texas. November 3–6, 1970.

Peterson, Arnold, and Ervin E. Gross, Jr. *Handbook of Noise Measurements.* West Concord, Mass.: General Radio Co., 1967.

Robinson, D. W. "The Concept of Noise Pollution Level." Report 38, Aerodynamics Division, National Physical Laboratory, March, 1969. Mimeographed.

Schneider, A. J. "Microphone Orientation in the Sound Field," *Sound and Vibration,* 4:2, February, 1970.

United States of America Standards Institute, *Acoustical Terminology.* S1.1-1960, New York, 1960.

ECONOMICS

Anon. "The Concorde's Boom Cut $3,120 Swath on Britain's Coast," *New York Times,* December 21, 1970.

Anon. Citizens League Against the Sonic Boom. "Correlation of Public Reaction to Sonic-Boom-Induced Property Damage." Fact Sheet 15, May, 1968.

Anon. "Noise Annoys," *Economist,* 213, December 19, 1964.

Anon. "Paying the Cost of Jet Aircraft Noise," *Engineering,* 199, March 19, 1965.

Dinenemann, P. F., and A. M. Lago. *External Costs and Benefits Analysis.* NECTP Report No. NECTP-224, U.S. Department of Transportation, Federal Railway Administration, Office of High Speed Ground Transportation, December, 1969.

Hayes, W. H., and P. M. Edge. "Effects of Sonic Boom and Other Shock Waves on Buildings," *Materials Research and Standards,* 4:11, November, 1964, pp. 588–593.

Hunter, George P. "Some Economic Aspects of the Aircraft Noise Problem." Paper presented at Industry/Government Symposium on STOL Transport Noise Certification, Washington, D.C., January 30, 1969.

McClure, P. T. "Indicators of the Effect of Jet Noise on the Value of Real Estate." Paper presented at the American Institute Aeronautics and Astronautics, 1969 Aircraft Design and Operations Meeting, Los Angeles, California, July 16, 1969.

——. *Some Projected Effects of Jet Noise on Residential Property Near Los Angeles International Airport by 1970.* Santa Monica, Calif.: Rand Corporation, April 1969.

Paik, I. K. *Impact of Transportation Noise on Urban Residential Property Values with Particular Reference to Aircraft Noise.* Washington, D.C., Report prepared for the Urban Mass Transportation Administration, August, 1970.

Ramsey, W. A. "Damage to Ottawa Terminal Building Produced by a Sonic Boom," *Materials Research and Standards,* 4:11, November, 1964, pp. 612–616.

Rummel, R. W. *Aircraft Noise Operational and Economic Considerations.*

President's Office of Science and Technology. Washington, D.C.: GPO, March, 1966, pp. 82–85.

U.S. Department of Interior, Report to the Secretary. "Sonic Boom Damage in the National Park System," January 10, 1967. Mimeographed.

U.S. Department of Transportation. "Summary of Sonic Boom Claims Presented in the US to the Air Force, Fiscal Years 1956–67." Washington, D.C., 1967. Mimeographed.

U.S. Federal Housing Administration. "Analysis of Residential Properties Near Airports." July, 1951. Mimeographed.

Waller, R. A. "Environmental Quality, Its Measurement and Control." Presented International Seminar on Urban Renewal, Brussels, Belgium, October 3–6, 1967.

BIBLIOGRAPHIES

British Ministry of Technology. *Noise Bibliography*. Vols, 1–7. London: HMSO, 1962–1968.

Highway Research Board. *Transportation Noise Bulletin*. Washington, D.C.: National Academy of Science, 1971.

Loring, J. C. G. *Selected Bibliography on the Effects of High-Intensity Noise on Man*. Monograph Supplement No. 2, *The Journal of Speech and Hearing Disorders*, 1953, American Speech and Hearing Association, Washington, D.C.

National Library of Medicine. *Effects of Noise on Man*. (193 references) Washington, D.C.: GPO, 1964–1968.

National Swedish Institute for Building Research. *Building Climatology. List of Literature. Part IV: Noise*. April, 1968 (779 references). PB 182 401, U.S. Department of Commerce.

U.S. Federal Aviation Agency/Library Services Division. "Aircraft Noise and Sonic Boom." Selected References. Bibliographic List No. 13, October, 1966 (460 references). Mimeographed.

Law
GENERAL

Anon. Statutory Instrument 1966, No. 424, "London (Heathrow) Airport Noise, Insulation Grants Scheme." London: HMSO, 1966.

Altree, Lillian R. Legal Aspects of Airport Noise and Sonic Booms, Part I, Chapters I–VII, prepared for the FAA, February, 1968.

Beranek, Leo L. "Technical and Legal Factors Concerning Jet Aircraft Noise." Presented at the National Conference on Noise as a Public Hazard, U.S. Public Health Service, Washington, D.C., June, 1968.

Bradley, William R. "Living with Noise Laws and Regulations," *Industrial Hygiene Foundation: Transactions Bulletin #40*, Proceedings 31st Meeting, 1966.

Conservation Foundation. "An Opening Attack on the Decibel Din: New Law Seeks to Control Aircraft Noise." *Conservation Foundation Newsletter*, August 30, 1968.

Goldstein, S. "Legal and Practical Limitations on Noise Control Methods." Presented at the International Conference on Reduction of Noise and Disturbance Caused by Civil Aircraft, London, November, 1966, INC/C4/P22.

Goodfriend, L. S., *et al*. "Social and Legal Implications of Noise Performance Regulations." Presented at the 80th Meeting Acoustical Society of America, Houston, Texas. November 3–6, 1970.

Grad, Frank P. "The Regulation of Occupancy, Noise, and Environmental Congestion: Selected Legal Aspects. Prepared under Contract PH 86-68-154 between the American Public Health Association and the Environmental Control Administration, Department of Health, Education and Welfare, 1970. Mimeographed.

Hatfield, Mark (U.S. Senator). "Compilation of State and Local Ordinances on Noise Control," *Congressional Record,* 115:176, 29, October, 1969, pp. E9031-E9112.

Katz, Milton. "The Function of Tort Liability in Technology Assessment," *University of Cincinnati Law Review,* 38, 1969, p. 587.

Marks, Laurence. "Police Chiefs Back-Pedal on Noise Meters," *London Observer,* September 22, 1968.

Masotti, Louis H., and Bruce Selton. "Aesthetic Zoning and the Police Power," *Journal of Urban Law,* 46, 1969, p. 773.

Miller, Arthur S. "Legal Responsibility for Anticipating Effects," a paper prepared for a symposium on "Extra-auditory Physiological Effects of Audible Sound," at the annual meeting of the American Association for the Advancement of Science, Boston, December 30, 1969.

Morris, John D. "Consumer Groups Seek Pre-Holiday Ban on Toys," *New York Times,* November 19, 1970.

National Research Council of Canada, Division of Applied Physics. *A Brief Study of a Rational Approach to Legislative Control of Noise.* Ottawa, Canada, 1968.

Odell, A. H. "Development and Enforcement of Noise Standards." Presented at International Conference on Reduction of Noise and Disturbance Caused by Civil Aircraft, London, November, 1966. Report No. INC/C4/P7.

Seago, Erwin. "The Airport Noise Problem and Airport Zoning," *Maryland Law Review,* 28:120, 1968.

Shenker, Israel. "Environmental Suits Crowd Courts," *New York Times,* November 15, 1969.

Spater, George A. "Noise and the Law," *Michigan Law Review,* 63:1373, June, 1965.

Tondell, L. M., Jr. "Noise Litigation at Public Airports," in *Alleviation of Jet Aircraft Noise Near Airports: A Report of the Jet Aircraft Noise Panel.* Washington, D.C.: GPO, March, 1966.

NATIONAL

Federal Aviation Administration. Civil Airplane Noise Reduction Retrofit Requirements Advance Notice of Propose/Rule Marking (Docket No. 10664; Notice No. 74-44) *Federal Register,* 35, November 4, 1970.

Federal Aviation Administration. Civil Aircraft Sonic Boom, Notice of Proposed Rule Making (Docket No. 10261; Notice No. 70-16) *Federal Register,* 35:74, April 16, 1970.

Goldstein, S. "A Problem in Federalism, Property Rights First in Airspace and Technology," in *Alleviation of Jet Aircraft Noise Near Airports: A Report of the Jet Aircraft Noise Panel.* Washington, D.C.: Government Printing Office, March, 1960, pp. 132–142.

Golovin, Nicholas E. "The Public and National Noise Standards." Presented at the 76th Meeting of the Acoustical Society of America, Cleveland, O., November 21, 1968.

Kramen, James M. "Noise Control: Traditional Remedies and a Proposal for Federal Action," *Harvard Journal on Legislation*, 7, 1970, pp. 533–566.

Labovitz, John R. "Federal Regulation of Air Transportation and the Environmental Impact Problem," *University of Chicago Law Review*, 35, 1968, p. 317.

Lindsey, Robert. "FAA Acts to Cut Noise of Jetliners," *New York Times*, November 13, 1969.

U.S. Congress. Aircraft Noise Control, Public Law 90-411,82 Federal Aviation Administration maximum noise standards and noise objectives, limiting the noise permissible for all new subsonic transport airplane types, January 3, 1969.

U.S. Congress. Clean Air Act, Title IV, Noise Pollution, Public Law 91-604, December 31, 1970.

U.S. Congress. Federal Aviation Act of 1958. S.R. No. 1811, 85th Congress, 2nd Session, 1958.

U.S. Congress. National Environmental Policy Act of 1969, Public Law 91-190, 83 Stat. 852, January 1, 1970.

U.S. Congress, House, Housing and Urban Development Act of 1965, H.R. 7984, July, 1965. Especially "Study Concerning Relief of Homeowners in Proximity to Airports," Sec. 1113.

U.S. Department of Health, Education and Welfare, "Hazardous Substances: Definitions and Procedural and Interpretative Regulations," Part 191, Chapter 1, Title 21, Food and Drugs, *Federal Register*, 35:246, December 19, 1970.

U.S. Federal Aviation Administration. Noise Standards: Aircraft Type Certification, Part 36, Chapter 1, Title 14—Aeronautics and Space. Issued January 3, 1969.

Walsh-Healey Public Contracts Act, Part 50-204, Safety and Health Standards for Federal Supply Contracts, Section 50-204.10 Occupational Noise Exposure, *Federal Register*, 34:96, May 20, 1969.

Anon. "California Board Votes Noise Curb, 15 Year Program is Adopted   STATE for Airports in State," *New York Times*, November 12, 1970.

Anon. "Noise Bureau to Conduct Studies, Submit Standards to State Control Board," *Environmental Reporter* 1:38, January 15, 1971, p. 100.

California, State of. Motor Vehicle Code, Divisions 11 and 12, paragraphs 23130 and 27160 (1968). Chapter 2, Title 13, *California Administrative Code*.

California, State of. "Progress Report," Environmental Quality Study Council, February, 1970.

California, State of. Senate Bill 2434, 1970, Regular Session.

California, State of. Senate Bill 1300, Introduced by Senator Beilenson, April 3, 1970.

Crotti, J. R. Report to the Legislature on Proposed Noise Standards for California Airports Pursuant to Public Utilities Code 21669, California Department of Aeronautics, April 1, 1970.

Heath, Warren. "California Experience in Vehicle Noise Enforcement."

Paper presented to Committee on Hearing, Bioacoustics and Biomechanics, National Academy of Science-National Research Council, April 1, 1971.

King, Seth S. "Snowbelt State Press to Regulate Snowmobiles," *New York Times,* March 2, 1971.

Little, Ross. "California Laws and Regulations Relating to Motor Vehicle Noise." Presented at the 78th Meeting of the Acoustical Society of America, San Diego, California, November 4, 1969.

Minnesota, State of. "Airport Zoning Act," Chapter III, 1969 Session Laws, St. Paul, Minnesota.

New York State Thruway. Motor Vehicle Noise Limit, Section 386, Vehicle and Traffic Law, October 1, 1965.

LOCAL

Anon. "Berkeley Council Rejects Copter Patrols By Police," *New York Times,* May 31, 1970.

Anon. "Families Ask $400,000 for Jet Noise" (S/V News), *Sound and Vibration,* 1:12, December, 1967.

Air-Conditioning and Refrigeration Institute, "Draft of a Municipal Ordinance to Regulate Sound," 1966.

Allegheny Airlines, Inc., v. Village of Cederhurst, 132 F. Supp. 871 (EDNY 1955), aff'd 238F. 2nd 812 (2nd Cir. 1968).

American Airlines, Inc., v. Audobon Park, 297 F. Supp. 207 (W.D. Ky. 1968).

American Airlines, Inc. v. Town of Hempstead, 272 F. Supp. 226 (EDNY 1966) aff'd 398 F. 2nd 369 (2nd Cir. 1968), cert. den., 393, U.S. 1017 (1969).

Bird, David. "Rickles Asks Noise Limits to Ban All Jets," *New York Times,* December 30, 1970.

Blazier, Warren E., Jr. "Criteria for Control of Community Noise," *Sound and Vibration,* 2:5, May, 1968.

Bolt, Beranek and Newman, Inc. *Noise Level Restrictions for the City of San Diego Industrial Park.* BBN Report No. 1270, February, 1966, pp. 8–9.

Chicago, City of Chicago Zoning Ordinance, 1969. Revised July, 1971.

Dietrich, Charles. "Preparation of Community Noise Ordinances." Paper presented to Committee on Hearing, Bioacoustics and Biomechanics, National Academy of Science–National Research Council, April 1, 1971.

Donley, Ray. "Community Noise Regulation," *Sound and Vibration,* 3:2, February, 1969, pp. 12–21.

Fredikson, H. M. "Noise Control on the Local Level." Paper presented at the American Medical Association's 6th Congress on Environmental Health, Chicago, April 28–29, 1969.

Fried, Joseph, P. "Revised Building Code Approved by City Council," *New York Times,* October 23, 1968.

Hanover, Township of, v. Morristown, Town of. New Jersey Superior Court, Chancery Division, December 10, 1969.

Hurlburt, Randall. "Noise Control Experience in Local Government." Paper presented to Committee on Hearing, Bioacoustics and Biomechanics, National Academy of Science-National Research Council, April 1, 1971.

New York City Building Code, Section 1208.0, Noise Control in Multiple Dwellings, December 6, 1968.

Odell, Albert H. "Jet Noise at John F. Kennedy International Airport." Presented before the Panel on Jet Aircraft Noise, Washington, D.C., October 29, 1965.

American Academy of Opthalomology and Otolaryngology. *Subcommittee on Noise in Industry: Guide for Conservation of Hearing in Noise.* Los Angeles: American Academy of Opthalomology and Otolargyngology, revised 1969.

STANDARDS

American National Standards Institute, Inc. New York City. (ANSI). S1.1-1960, Acoustical Terminology (including Mechanical Shock and Vibration).

——. S1.2-1962, *Method for Physical Measurement of Sound.*

——. S1.4-1961, *Specification for General-Purpose Sound Level Meters.*

——. S1.6-1967, *Preferred Frequencies and Band Numbers for Acoustical Measurements.*

——. S1.10-1966, *Method for the Calibration of Microphones.*

——. S1.11-1966, *Specification for Octave, Half-Octave, and Third-Octave Band Filter Sets.*

——. S1.12-1967, *Specifications for Laboratory Standard Microphones.*

——. S3.4-1968, *Procedure for the Computation of the Loudness of Noise.*

——. S3-W39, *The Effects of Shock and Vibration on Man.*

——. Y10.11-1953 (R1959) *Letter Symbols for Acoustics.*

——. Z24.15-1955, *Method for Specifying the Characteristics of Analyzers Used for the Analysis of Sound and Vibrations.*

——. Z24.19-1957, *Practice for Laboratory Measurement of Air-borne Sound Transmission Loss of Building Floors and Walls.*

——. Z24.X2, *The Relations of Hearing Loss to Noise Exposure.*

German Engineers Association (VDI). *Evaluation of Working Noise in the Neighborhood,* VDI Regulation 2058, Parts 1 and 2, August, 1968.

Goromosov, M. S. *The Physiological Basis of Health Standards for Dwellings,* Public Health Paper #33. Geneva: World Health Organization, 1968.

International Organization for Standardization (ISO). R131-1959, *Expression of the Physical and Subjective Magnitudes of Sound and Noise* (Agrees with S1.1-1960). Including Supplement R357-1963.

——. R140-1960. *Field and Laboratory Measurements of Air-Borne and Impact Sound Transmission.*

——. R226-1961, *Normal Equal-Loudness Contours for Pure Tone and Normal Threshold of Hearing under Free Field Listening Conditions.*

——. R357-1963, *Expression of Power and Intensity Levels of Sound or Noise.* Supplement to P131-1959.

——. R362-1964, *Measurement of Noise Emitted by Vehicles.*

——. R454-1965, *Relation between Sound Pressure Levels of Narrow Bands of Noise in a Diffuse Field and in a Frontally-Incident Free Field for Equal Loudness.*

——. R495-1966, *General Requirements for the Preparation of Test Codes for Measuring the Noise Emitted by Machines.*

——. R507-1966, *Procedure for Describing Aircraft Noise around an Airport.*

ISO/TC 43, *Noise Assessment with Respect to Community Noise.* Draft Recommendation, May 15, 1970.

North Atlantic Treaty Organization. *NATO Standard Agreement.* STANAG 3473 Hearing Conservation, January 14, 1969.

Ponomarenko, I. I. "Standards of High Frequency Industrial Noise for Young Workers," *Russian Industrial and Hygiene Sanitation,* Translation 1969.

Society of Automotive Engineers, Inc. (SAE). J6A, *Ride and Vibration Data Manual.* New York City.

——. J919, *Measurement of Sound Level at Operator Station.*

——. J952a, *Sound Levels for Engine Powered Equipment.*

——. J331, *Sound Levels for Motorcycles and Motor Driven Cycles.*

——. J336, *Sound Level for Truck Cab Interior.*

——. J366, *Exterior Sound Level for Heavy Trucks and Buses.*

——. J672a, *Exterior Loudness Evaluation of Heavy Trucks and Buses.*

——. J986a, *Sound Level for Passenger Cars and Light Trucks.*

——. J994, *Criteria for Backup Alarm Devices.*

U.S. Air Force. *Hazardous Noise Exposure,* AFR 160-3A, June 27, 1960.

U.S. Army, Army Material Command. *Maximum Noise Level for Army Materiel Command Equipment,* HEL Standard S-1-63B, Human Engineering Laboratories, Aberdeen Proving Ground, Maryland, June, 1969 (to be revised 1971).

U.S. Department of the Army. Technical Bulletin, *Noise and Conservation of Hearing,* TB MED 251, January, 1965 (to be revised, 1971).

U.S. Department of the Navy. *Outline of Hearing Conservation Program,* BUMED INST 6260.6B, March, 1970.

Noise Abatement
GENERAL

Anon. "AOCI Withdrawal from NANAC." *Newsletter, National Aircraft Noise Abatement Council,* 9:2, February 15, 1968.

Anon. "Dynamic Measurement Instrumentation Buyer's Guide" *Sound and Vibration,* 4:12, December, 1970.

Anon. National Noise Standards Proposed in Draft Legislation by Administration, *Environmental Reporter* 1:44, February 26, 1971.

Anon. "Noise and Vibration Control Systems," *Sound and Vibration,* 4:7, July, 1970.

Anon. "Noise and Vibration Control Systems," *Sound and Vibration,* 4:8, August, 1970.

Anon. "The War against Runaway Decibels," *Saturday Review,* March 7, 1970, p. 54.

Airport Operators Council, Inc. *AOCI Airport Operator's Aircraft Noise Kit.* Washington, D.C., August, 1967.

——. "An Analysis of Aircraft Noise Annoyance and Some Suggested Solutions." March, 1960. Mimeographed.

Bird, David. "Environmental Superagency Asks City for Half Billion for Projects," *New York Times,* October 30, 1968.

Beranek, Leo. "The Effects of Noise on Man and Its Control." Presented at the Atmospheric Noise Pollution and Measures for Its Control Course, Berkeley, California, June 17–21, 1968.

——. *Noise Reduction.* New York: McGraw-Hill, 1960.

Bolt, Beranek and Newman, Inc. *Analysis of Community and Airport Relationships/Noise Abatement.* BBN Report No. 1254, December, 1965.

Bragdon, Clifford R. "Noise Pollution: The Problem and Its Control." Presented at the 31st Annual Planning Institute, 1969, Albany, N.Y., New York Planning Federation, February, 1970.

Braun, Fred W. "Hearing Conservation in Industry," *Noise Control,* 4:7, July, 1958.

Bruce, David D., and Parker W. Hirtle. "Noise Control in Home Heating and Air-Conditioning Systems." Presented at the 80th Meeting Acoustical Society of America, Houston, Texas. November 3–6, 1970.

Burrows, William E. "Hush-Hush Agent Helps Airlines Beat Noise Ban," *New York Times,* October 20, 1967.

Collis, R. T. H. "Treating the Sonic Boom," *Astronautics and Aeronautics,* 8:6, 1970, pp. 42–43.

Council of Europe. *Noise Abatement: A Public Health Problem.* Luxembourg, 1965.

Cousins, Norman. "New York's Fight against Pollution," *Saturday Review,* March 7, 1970, p. 53.

Cuadra, E., and R. D. Beland. "Rationale for the Comprehensive Control of Urban Noise." Paper presented at the 16th annual meeting of the Institute of Environmental Sciences: "The Environmental Challenge of the Seventies," Boston, April, 1970.

Dygert, P. K. "A Public Enterprise Approach to Jet Aircraft Noise Around Airports," in *Alleviation of Jet Aircraft Noise Near Airports: A Report of the Jet Aircraft Noise Panel.* Washington, D.C.: GPO, March, 1966, pp. 107–116.

Farnsworth, Clyde H. "Conference on Sonic Boom Told Noise Can't Be Designed Away," *New York Times,* February 4, 1970.

Federal Aviation Agency, Office of Policy Development. *Allocating the Cost of New Programs to Alleviate Aircraft Noise Near Airports.* February, 1967.

Fellows, Lawrence. "Germans Who Complained of Airport Noise Get a Brand New Town," *New York Times,* March 14, 1971.

France. Ministry for Industrial and Scientific Development. "Nuisance Control in the City." First report on Vaudrevil, France, April, 1970 (English Abstract, U.S. Department of Housing and Urban Development). "Pollution Preventive in Cities" (Main Recommendations) April, 1970 (Translation, mimeographed, U.S. Department of Housing and Urban Development).

Gansberg, Martin. "Grants Will Aid Pollution Study," *New York Times,* July 6, 1970.

Globe, Emerson. "Lip Service to Noise Control," *Architectural Record,* 134:11, November, 1964.

Golovin, N. E. "Alleviation of Aircraft Noise," *Astronautics and Aeronautics,* 5:1, 1967, pp. 71–75.

Goodfriend, Lewis S. "Control of Noise through Propaganda and Education," *Noise as a Public Health Hazard.* ASHA Reports 4, The American Speech and Hearing Association, Washington, D.C., February, 1969, pp. 342–346.

Grootenhuis, P. "The Attenuation of Noise and Ground Vibrations from Railways," *Journal of Environmental Sciences,* 10:2, 1967, pp. 14–19.

Harr, Charles. "Airport Noise and the Urban Dweller: A Proposed Solution," Speech, Practicing Law Institute, New York, May 10, 1968.

Harris, Cyril M. *Handbook of Noise Control.* New York: McGraw-Hill, 1957.

Hubbard, H. H., D. J. Maglieri, and W. L. Copeland. "Research Approaches to Alleviation of Airport Community Noise," *Journal of Sound and Vibration,* 5:2, 1967, pp. 377–390.

Inglewood, California. *The Ten Point Action Program for the Alleviation of Noise Pollution in Inglewood California.* January 1, 1970.

Joiner, John W. "State of the Art in Residential Noise Control." Presented at the 80th Meeting Acoustical Society of America, Houston, Texas. November 3–6, 1970.

Lees, A. "New Ideas for Noise Control at Home," *Popular Science,* 197:3, September, 1970.

McVey, G. F. "Designing for Quiet," *Design & Environment,* 1:4, Winter, 1970.

Mecklin, H. "Its Time to Turn down All That Noise," *Fortune,* October, 1969.

Metzler, Dwight F. "Quiet Communities for New York." Presented at annual meeting American Public Health Association, October 29, 1971.

National Aeronautics and Space Administration. "Progress of NASA Research Relating to Noise Alleviation of Large Subsonic Jet Aircraft." A conference held at Langley Research Center, NASA SP-189, Washington, D.C., October, 1968.

——. *Progress of NASA Research Relating to Noise Alleviation of Large Subsonic Jet Aircraft.* NASA SP-189, 1968.

Naughton, James M. "Nixon Proposes 2 New Agencies on Environment," *New York Times,* July 10, 1970.

Newman, Robert B. "Household Noise, Can Anything Be Done about It?" Presented at the 80th Meeting Acoustical Society of America, Houston, Texas. November 3–6, 1970.

New York City, Mayor's Task Force on Noise Control. *Toward a Quieter City.* Report, New York City, 1970.

Odell, A. H. "Jet Noise at John F. Kennedy International Airport," in *Alleviation of Jet Aircraft Noise Near Airports. A Report of the Jet Aircraft Noise Panel.* Executive Office of the President. Washington, D.C.: Government Printing Office, March, 1966.

Paullin, Robert L. "The Federal Aviation Administration and Aircraft Noise Control." Presented at Seminar on Atmospheric Noise Pollution and Measures for Its Control, University of California, Berkeley, Calif., June 17–21, 1968.

Paullin, R. L., and J. F. Miller. "Aircraft Noise Abatement—The Prospects for a Quieter Metropolitan Environment." AIAA Paper No. 69-800, New York, 1969. Mimeographed.

Petts, E. C. "Practical Noise Control at International Airports with Special Reference to Heathrow," *Royal Aeronautics Society Journal,* 70:672, December, 1966, pp. 1051–1060.

Richter, G., and R. Hoch. "Problems of Noise around Airports and Means of Reducing It." Association Française des Ingenieurs et Techniciens de l'Aeronautique et de l'Espace, 8th Congress International Aeronautique, Paris, France, May 29–31, 1967. (In French.)

Rose, Thomas H. "Airport Noise Zoning Ordinance and Building Code." Presented at the 80th Meeting Acoustical Society of America, Houston, Texas. November 3–6, 1970.

Serendipity, Inc. *A Study of Magnitude of Transportation Noise Generation and Potential Abatement.* 3rd Quarterly Report, II: Serendipity, Inc., prepared for the Office of Noise Abatement, Department of Transportation, Washington, D.C., March, 1970.

Stephenson, R. J., and G. H. Vulkan. "Urban Planning against Noise," *Official Architecture and Planning,* 30:5, May, 1967.

Stevenson, G. M. "To Insure Domestic Tranquility: An Analysis of the Political Subsystem of Jet Aircraft Noise Abatement." Unpublished dissertation, Columbia University, 1970.

U.S. Department of Housing and Urban Development. "Noise Abatement and Control: Departmental Policy, Implementation Responsibilities, and Standards." Circular 1390.2, August 4, 1971.

U.S. Department of Transportation. *First Federal Aircraft Noise Abatement Plan FY 1969–70.* Washington, D.C., November, 1969.

U.S. House of Representatives, Committee on Interstate and Foreign Commerce. *Aircraft Noise Abatement Hearing.* Washington, D.C.: Government Printing Office, 1968. Statement by Hon. Byron G. Rogers, Representative, Colorado.

——. Office of Noise Abatement. "Summary Status Report Federal Aircraft Noise Abatement Program." Washington, D.C., April, 1968. Mimeographed.

U.S. Office of Science and Technology. "Alleviation of Jet Aircraft Noise Near Airports." A Report of the Jet Aircraft Noise Panel. Washington, D.C.: GPO, 1966.

U.S. Executive Office of the President, Bureau of the Budget, Office of Management and Budget. *Federal Budget for Fiscal Year 1971* (preliminary data on air, water, and noise pollution control programs, August 6, 1970).

——. *Federal Environmental Programs.* January, 1971.

Veneklasen, Paul S. "Community Noise Control," *Noise as a Public Health Hazard.* Washington, D.C.: American Speech and Hearing Association, 1969.

Wald, B. G., and P. H. Tope. "Household Noise Problems—Appliances." Presented at the 80th Meeting Acoustical Society of America, Houston, Texas. November 3–6, 1970.

NOISE SOURCE

Anon. "FAA to Offer Antinoise Plan for 2,000 Airliners Next Month," *New York Times,* October 12, 1970.

Anon. "The First Quiet Portable Compressor," *Sound and Vibration,* 3:5, May, 1969.

Anon. "Quieter and More Economical Hovercraft," *Engineering,* 205:5313, 1968, pp. 248–249.

Beranek, Leo, and Rudmose, R. Wayne. "Sound Control in Airplanes," *Journal of the Acoustical Society of America,* 19:2, March, 1947.

Botsford, James H. "Control of Industrial Noise through Engineering," *Noise as a Public Health Hazard.* Washington, D.C.: American Speech and Hearing Association, 1969.

——. "Engineering Standards and Specifications." Lecture presented Pennsylvania State University, College Station, Pennsylvania. July, 1967.

Bredin, Howard W. "City Noise: Designers Can Restore Quiet at a Price," *Product Engineering*, 39:24, November 18, 1968.

Chanaud, Robert C. "Noise Reduction of a Household Refrigerator." Presented at the 80th Meeting Acoustical Society of America, Houston, Texas. November 3–6, 1970.

General Motors Corp. *Operation Hush: A Synopsis of the Commercial Vehicle Noise Reduction Programs by the General Motors Truck and Coach Division*, June, 1970.

Levin, S. M. "Aircraft Noise—Can It Be Cut?" *Space/Aeronautics*, 46:2, 1966, pp. 65–75.

Lindsey, Robert. "A Plan to Muffle All Jets Draws Airline Opposition," *New York Times*, February 14, 1971.

Manning, G. P. "Down with Decibels: Part I," *Motor Boating*, February, 1967, pp. 34–38.

——. "Down with Decibels, Part II: Practical Ways of Quieting Engine Noise," *Motor Boating*, March 1967, pp. 46–49.

Marsh, A. H. "Study of Acoustical Treatment for Jet-Engine Nacelles," *The Journal of the Acoustical Society of America*, 43:5, 1968, pp. 1137–1146.

Meyersburg, B. B., and C. H. Williams. "The Two-Segment Noise Abatement Approach to Landing." Presented at International Conference on the Reduction of Noise and Disturbance Caused by Civil Aircraft, Committee No. 3, London, November, 1966.

Ortega, J. C. *Diesel Engine Horn Sound Level Reduction Study for the City of Inglewood, California*, May 16, 1969.

Simpson, Wayne C. "How to Develop Purchase Safeguards for Noise Control," *National Safety News*, August, 1970.

NOISE PATH

Anon. "Control of Noise by the Siting of Aerodomes, Planning Their Layout and Limiting Residential Development Nearby." Presented at International Conference on Reduction of Noise and Disturbance Caused by Civil Aircraft, London, November, 1966. Report No. INC/C2/P2.

Arde, Inc., and Town and City, Inc. *A Study of the Optimum Use of Land Exposed to Aircraft Landing and Takeoff Noise*. Prepared for NASA, March 1966, NASA CR 410.

Beaton, John L., and L. Bourget. "Can Noise Radiation from Highways be Reduced by Design." Presented at the 47th Annual Meeting of the Highway Research Board, January, 1968.

Cohen, Alexander. "Location-Design Control of Transportation Noise," *Journal of the Urban Planning and Development Division: ASCE*, 93:4, December, 1967.

Embleton, T. F. W. "Sound Propagation in Homogeneous Decidious and Evergreen Woods." *Journal of the Acoustical Society of America*, 35:8, August, 1963.

Glaser, E. M. "On Narrow-band FM Noise ASA Masking Noise," *Journal of the Acoustical Society of America*, 47:2, February, 1970.

Gordon, C. G. "Land-Use Planning and Industrial Noise." Paper 6B5 presented at 71st Acoustical Society of America Meeting, Boston, Mass., June, 1966. Also in *Journal of the Acoustical Society of America*, 39:6, 1966, p. 1255.

Kaufman, James J. "Effective Noise Control through Planning." Presented at the American Medical Association's Sixth Congress on Environmental Health, Chicago, 28–29, April, 1969.

Kugler, B. A., C. G. Gordon, and W. J. Gallowary. "Highway Noise: a Design Guide for Highway Engineers." Unpublished report for the National Cooperative Highway Research Program No. NCHRP 3–17, June, 1970.

Kurze, U. J. "Prediction of Sound Attenuation by Barriers." Presented at the 80th Meeting Acoustical Society of America, Houston, Texas. November 3–6, 1970.

McGrath, Dorn C. "Compatible Land Use." Presented at the ASCE/AOCI Joint Specialty Conference on Airport Terminal Facilities, Houston, Texas, April 14, 1967.

———. "Planning for Compatible Land Use around Airports." Presented at the National Conference of the American Association of Airport Executives, Philadelphia, Pa., May 22, 1968.

Northwood, T. D., *et al. Noise Control in Residential Buildings.* Division of Building Research, National Research Council, Ottawa, Canada, February, 1967.

Sawyer, F. L. "Aircraft Noise and Siting of a Major Airport," *Journal of Sound and Vibration*, 5:2, 1967, pp. 355–363.

Strunch, J. E. "An Analysis of the Advantages and Difficulties of Zoning Regulations for Chicago—O'Hare International Airport," in *Alleviation of Jet Aircraft Noise Near Airports: A Report of the Jet Aircraft Noise Panel.* Washington, D.C.: GPO, March, 1966, pp. 151–156.

Sullivan, T. M. "Criteria for the Location of Airport Sites." Presented at International Air Transport Association, 17th Technical Conference, Lucerne, Switzerland, October 9–14, 1967. (Montreal: ITA, 1968.)

United Kingdom, Minister of State, Board of Trade. *Aircraft Noise.* Report of an International Conference on the Reduction of Noise and Disturbance Caused by Civil Aircraft. London: HMSO, 1967.

U.S. Department of Housing and Urban Development. *Airport Environs: Land Use Controls: An Environmental Planning Paper.* Washington, D.C., May, 1970.

U.S. Office of the White House. "Aircraft Noise and Compatible Land Use in the Vicinity of Airports." Memorandum for Heads of Departments and Agencies, March 22, 1967.

Veneklasen, P. S. "Use of Building Barriers for Airport Neighborhood Noise Control." Presented at 72nd Acoustical Society of America Meeting, Los Angeles, California, November, 1966, paper 3D8. Also in *Journal Acoustical Society of America*, 40:5, 1966, p. 1254.

Anon. "City Council Approves Revenue Bond Issue and Authorized **OBJECT** Demonstration Program for Soundproofing," *Newsletter, National Aircraft Noise Abatement Council*, 8:8, August 15, 1967.

Anon. "Eighty-six Airports Noise Grants in Six Months," *London Daily Telegraph,* December 10, 1966.

Anon. "Few Seeking Grants to Cut Air Noise," *London Daily Telegraph,* March 10, 1968.

Anon. "Hearing Protective Devices," *Sound and Vibration,* 4:11, November, 1970.

Anon. "New Sound Insulation Techniques Applied to a Motel Near Chicago's O'Hare Airport." *Newsletter, National Aircraft Noise Abatement Council,* 9:9, September 15, 1968.

Bishop, Dwight E. "Reduction of Aircraft Noise Measured in Several School, Motel, and Residential Rooms," *Journal of the Acoustical Society of America,* 39:5, May, 1966.

Bolt, Beranek and Newman, Inc. "Design Guide: Methods for Improving the Noise Insulation of Houses with Respect to Aircraft Noise." Prepared under contract for the Federal Housing Administration, Department of Housing and Urban Development, November, 1966. Mimeographed.

Eijk, J. Van Den. "The New Dutch Code on Noise Control and Sound Insulation in Dwellings and Its Background," *Journal of Sound and Vibration,* 3:1, 1966.

Itow, T. "A Procedure for Reducing Noise in School Rooms Near an Airport," Paper F-3-1, in *Proceedings of the 6th International Conference on Acoustics.* Tokyo, August 21–28, 1968, Vol. IV, pp. F73–F76.

Lawrence, R. "The Influence of Traffic Noise on the Design of External Walls and Buildings." Paper presented at the 6th International Congress on Acoustics, Tokyo, Japan, 1968.

U.S. Department of Housing and Urban Development. *A Guide to Airborne, Impact, and Structure Borne Noise—Control in Multifamily Dwellings."* R. D. Berendt, G. E. Winzer and C. B. Burroughs, authors. Washington, D.C.: GPO, September, 1967.

Winzer, George. "Noise Control in Housing." Paper presented to Committee on Hearing, Bioacoustics and Biomechanics, National Academy of Science–National Research Council, April 1, 1971.

Wyle Laboratory. *Final Report on the Soundproofing Pilot Project for the Los Angeles Department of Airports.* Report No. WCR 70-1, Wyle Laboratory Research Staff, El Segundo, Calif., 1970.

Noise Sources
GENERAL

Bolt, Beranek and Newman, Inc. *Noise in Urban and Suburban Areas.* Washington, D.C.: GPO, January, 1967.

Bonvallet, G. L. "Levels and Spectra of Traffic, Industrial and Residential Noise," *Journal of the Acoustical Society of America,* 23:4, July, 1951, pp. 435–439.

Branch, Melville, C., et al. *Outdoor Noise and the Metropolitan Environment: Case Study of Los Angeles with Special Reference to Aircraft.* Los Angeles; UCLA, 1970.

Brown, Edward, et al. *City Noise.* New York City Department of Health, Noise Abatement Commission. New York: Academy Press, 1930.

Chalupnik, J. D., ed. *Transportation Noises: A Symposium on Acceptability Criteria.* Seattle: University of Washington Press, 1970.

Cook, B. D., *et al.* "Noise Survey of Houston, Texas." Presented at the 80th Meeting Acoustical Society of America, November 3–6, 1970.

Goodfriend, L. "Noise in This Community," *Noise Control,* 4, 1958, pp. 22–28.

Hardy, H. C. "Twenty-five Years' Research in Outdoor Noise," *Noise Control,* 1, January, 1965, pp. 20–24.

Ostergaard, P. B., and R. Donley. "Background Noise Levels in Suburban Communities," *Journal of the Acoustical Society of America,* 36:3, 1964, pp. 409–413.

Stevens, K. N. *A Survey of Background and Aircraft Noise in Communities near Airports.* NACA Technical Note 3379, December, 1954.

U.S. Congress, House, Subcommittee on Transportation and Aeronautics. *Hearings.* 90th Congress, 1st Session, H.R. 3400, November 14, 1967–March 20, 1968.

Veneklasen, Paul S. "City Noise—Los Angeles," *Noise Control.* 2:3, July, 1956.

Wilson Committee on the Problem of Noise. *Noise: Final Report.* London: HMSO, 1963.

U.S. Congress, House, Committee on Interstate and Foreign Commerce, *Aircraft Noise Abatement Hearings, November 14, 1967–March 20, 1968.* Washington, D.C.: GPO, 1968.

Anon. "Eight Hundred New Airports Needed, FAA Says," *Philadelphia Inquirer,* November 10, 1968.

Anon. "Helicopter Noise Blamed in Part for Two Deaths at Kennedy Train," *New York Times,* October 30, 1968.

Anon. "Noise Conference in Massachusetts," *Newsletter, National Aircraft Noise Abatement Council,* 19:7, July 15, 1968.

Anon. "Noise Generating Characteristics of Rolls-Royce RB.211 Engine," *Newsletter, National Aircraft Noise Abatement Council,* 19:4, April 15, 1968.

Anon. "Pan Am Will Run Heliport in City," *New York Times,* October 19, 1968.

Anon. "Senate Committee Trims $16 Million from SST Appropriation Request," *New York Times,* December 17, 1969.

Aakre, B. "Aircraft Noise Monitoring System in Use in Norway." Presented at International Conference on Reduction of Noise and Disturbance Caused by Civil Aircraft, 1967. In *Aircraft Noise.* London: HMSO, 1967.

Beranek, L. L. "General Aircraft Noise," in *Noise as a Public Health Hazard,* eds. W. D. Ward and J. E. Fricke. Washington, D.C.: American Speech and Hearing Association, February, 1969, pp. 256–269.

Bird, W. H., and T. L. Wilde. "Aircraft Noise and the Community," *Canadian Aeronautics and Space Journal,* 14:6, 1968, pp. 201–209.

Black, R. E. "DC-8 Stretched Jets' Should Produce Less Noise Annoyance at Airports," *SAE Journal,* 75:6, 1967, pp. 51–55.

Block, Victor. "The Supersonic Transport and You," *Science Digest,* 60:1, July, 1966.

Boeing. *Airport and Community Noise of the SST.* Boeing/General Electric, Seattle, March 1969.

AIR TRANSPORTATION

Bolt, Beranek and Newman, Inc. *A Discussion of Current and Potential Noise Problem Areas Influencing the Development of Civil Aviation.* Report No. 1618 prepared for Ad Hoc Committee on Aircraft Noise, Aeronautics and Space Engineering Board of the National Academy of Engineering, Washington, D.C., March, 1968.

Bruckmeyer, F., and J. Land. "Monitoring Aircraft Noise Around An Airport." Presented at International Conference on Reduction of Noise and Disturbance Caused by Civil Aircraft, London, November, 1966. In *Aircraft Noise.* London: HMSO, 1967.

Burrows, William E. "City Back Housing Despite Jet Fear," *New York Times,* October 26, 1967.

Clark, Evert. "Noise Called Bar to New Airports," *New York Times,* October 5, 1967.

Cole, John N., and Robert T. England. *Evaluation of Noise Problems Anticipated with Future VTOL Aircraft.* Aerospace Medical Research Laboratories, Wright-Patterson Air Force Base, Ohio, AMRL-TR-66-245, May, 1967.

Copeland, W. L., *et al. Noise Measurement Evaluations of Various Takeoff-Climbout Profiles of a Four-Engine Turbojet Transport Airplane.* NASA TN D-3715, 1966.

———. *Further Studies of Noise Produced by a Boeing 727 Turbofan Transport Airplane for Various Takeoff-Climbout Profiles.* NASA Langley Working Paper 385. March 15, 1967.

Detroit Metropolitan Area Regional Planning Commission. *Environs Study and Plan.* Detroit Metropolitan Wayne County Airport, May, 1964.

Finney, John. "Congress Closes as Senate Votes SST Compromise," *New York Times,* January 3, 1971.

Fish, E. B. "Flyover Noise during Takeoff and Landing of the DC-9-30 Aircraft." Douglas Aircraft Co., Inc., Long Beach, Calif., Report No. DAC-33753, May, 1967. N8-322221. Mimeographed.

Franken, P. A., and D. Standley. *Aircraft Noise and Airport Neighbors: A Study of Logan International Airport.* Bolt, Beranek and Newman, prepared for the Department of Transportation and the Department of Housing and Urban Development, Report No. DOT/HUD IANP-70-1. Washington, D.C., March 1970.

Gasaway, Donald C. "Aeromedical Significance of Noise Exposure Associated with the Operation of Fixed- and Rotary-Winged Aircraft." Brooks Air Force Base, Texas; U.S. Air Force School of Aerospace Medicine, November, 1965. Mimeographed.

Green, D. M. "Sonic Boom," *Psychology Today,* 2, 1968, pp. 38–44.

Hess, J. L. "French Investigate Deaths of Three Linked to Superjets' Boom," *New York Times,* August 3, 1967.

Hilton, D. C., A. C. Dibble, Jr., and W. L. Copeland. "Measurements of Noise Produced by a Boeing 727 Turbofan Transport Airplane During Takeoff-Climbout Operations." NASA Langley Working Paper 214, April 20, 1966. Mimeographed.

Hurdle, P., *et al.* "Jet Aircraft Noise in Metropolitan Los Angeles Under Air Route Corridors." Presented at the 80th Meeting Acoustical Society of America, Houston, Texas, November 3–6, 1970.

Kenworthy, E. W. "Memo List Boeing Failures on SST," *New York Times,* December 19, 1970.

Kryter, K. D. "Sonic Booms from Supersonic Transport," *Science* 163, December 19, 1970.

——. "Sonic Booms from Supersonic Transport," *Science* 163, January 23, 1969, pp. 359–367.

Lindsey, Robert. "Inflation Is Adding to SST Cost, According to an Aide of Boeing," *New York Times,* October 17, 1969.

Lydon, Christopher. "Funds for the SST Is Voted in House by a Slim Margin," *New York Times,* May 28, 1970.

——. "Higher Airport Noise Level Foreseen," *New York Times,* February 25, 1970.

——. "House Declines to Back Senate in SST Fund Curb," *New York Times,* December 9, 1970.

——. "Nixon Backs SST Project; Battle Looms in Congress," *New York Times,* September 24, 1969.

——. "Project Chief Lobbies Hard to Sell the SST," *New York Times,* July 20, 1970.

——. "Senate Approves Two Curbs on SST," *New York Times,* December 3, 1970.

——. "Senate Delays SST Plan Returning It to Conferees," *New York Times,* December 30, 1970.

——. "The Senate Defeats Motion to Close Debate over SST," *New York Times,* December 19, 1970.

——. "Senate SST Filibuster Pledged Conferees Will Convene Today," *New York Times,* December 10, 1970.

——. "Short Term Funding for SST Is Viewed as Futile," *New York Times,* December 25, 1970.

——. "Supporters of SST Waver in Senate," *New York Times,* August 28, 1970.

Lundberg, B. "The Menace of the Sonic Boom to Society and Civilization." Aeronautical Research Institute of Sweden, 1966. Mimeographed.

Lyster, H. N. C. "The Nature of the Sonic Boom," *Materiels Research and Standards,* 4:11, November, 1964, pp. 582–587.

Melinikov, B. N. "Noise Generated on the Ground During Takeoff and Landing of the Tu-124 Passenger Aircraft," *Soviet Physical Acoustics,* 11, October–December, 1965, pp. 170–172.

Great Britain, Ministry of Aviation. *Aviation Noise.* London: HMSO, undated.

National Academy of Science–National Research Council. *Sonic Boom Generation and Propogation.* Washington, D.C., June, 1968.

Nixon, C. W., *et al. Sonic Booms Resulting From Extremely Low-Altitude Supersonic Flight.* Aerospace Medical Research Laboratory, Wright-Patterson Air Force Base, Ohio, AMRL-TR-68-52, October, 1968.

Odell, Albert H. "Jet Noise at John F. Kennedy International Airport." Presented before the Panel on Jet Aircraft Noise, Washington, D.C., October 29, 1965.

Ossipov, G. L., and A. A. Klumuhin. "Measurement and Evaluation of

Aircraft Noise in Flight." Paper F-3-2, in *Proceedings of the 6th International Congress on Acoustics,* Tokyo, August 21–28, 1966, Vol. IV, pp. F77–F80.

Pattarini, C. B. "The Aviation Noise Problem," *Journal of Sound and Vibration,* 5:2, 1967, pp. 370–376.

Powers, John, and Kenneth Power. *The Supersonic Transport—The Sonic Boom and You.* FAA paper, Department of Transportation, Washington, D.C.

Satre, P. "Supersonic Air Transport True Problems and Misconceptions," *Journal of Aircraft,* 7:1, 1970, pp. 3–12.

Schmeck, Harold M., Jr. "Sonic Boom Acceptance Test Urged," *New York Times,* November 22, 1968.

Sherril, Robert. "The Jet Noise Is Getting Awful," *New York Times,* January 14, 1968.

Sikorsky. *VTOL, The Noise Question.* Sikorsky Aircraft Brochure, Stratford, Conn., 1970.

Simmons, Duane R. "Use of Acoustical Modelling in Airport-Noise Studies" (presented at 72nd Acoustical Society of America Meeting, November 1966, Los Angeles, California), *Journal of the Acoustical Society of America,* 40:5, 1966, p. 1254.

Tracor, Inc. *Study of Community Response to Aircraft Noise and Sonic Boom.* Prepared for NASA, 1970.

United Kingdom, Board of Trade. "Aircraft Noise." Report of an International Conference on the Reduction of Noise and Disturbance Caused by Civil Aircraft, Lancaster House, London, November 22–30, 1966. London: HMSO, 1967.

U.S. Congress, House, Committee on Interstate and Foreign Commerce. Aircraft Noise Problems, *Hearings,* September 7, 1959–December 6, 1962. Washington, D.C.: GPO, 1963.

U.S. Department of Interior. "Report of the Secretary of the Interior of the Special Study Group on Noise and Sonic Boom in Relation to Man." November, 1968. Mimeographed.

U.S. White House. "Text of Statement by President Nixon on SST," *New York Times,* December 6, 1970.

Voce, J. E. "Noise from Aircraft," *Engineering,* 204:5303, 1967, pp. 983–986.

Witkin, Richard. "Senate Rejects SST Fund in 52–41 Vote after Drive by Environmental Lobby," *New York Times,* December 4, 1970.

Zimmerman, Fred L. "Supersonic Snow Job," *Wall Street Journal,* February 9, 1967.

GROUND
TRANSPORTATION

Apps, David. "Cars, Trucks and Tractors as Noise Sources," *Noise as a Public Health Hazard.* Washington, D.C.: American Speech and Hearing Association, 1969.

Bender, E. K., and M. Heckl. *Noise Generated by Subways and in Stations,* Report No. OST-ONA-7-0-1, Office of Noise Abatement, Department of Transportation, Washington, D.C., January, 1970.

Bottom, C. G., and D. J. Croome. "Road Traffic Noise—Its Nuisance Value," *Applied Acoustics,* 2:4, 1969, pp. 279–296.

Callaway, D. B., and A. H. Hall. "Laboratory Evaluation of Field Measurements of the Loudness of a Truck Exhaust Noise," *Journal of the Acoustical Society of America,* 26:1, 1954, p. 140. (Abstract)

Canada, National Research Council. *Snowmobile Noise: Its Sources, Hazards and Control,* APS-477, 1970.

Consumers Union. "The Little Cars," *Consumer Reports,* 36:1, January, 1971.

Dietrich, C. W., *et al. High Speed Ground Transportation Noise.* Bolt, Beranek and Newman, prepared for TRW Systems, Report No. 1741, Washington, D.C., October, 1968.

Embleton, T. F. W., and G. J. Thiessen. "Train Noise and Use of Adjacent Land," *Sound* 1:1, January–February, 1962, pp. 10–16.

Galloway, W. J., W. E. Clark, and J. S. Kerrick. Highway Noise: Measurement, Simulation, and Mixed Reactions. Bolt, Beranek and Newman, National Cooperative Highway Research Program, Report 78, Washington, D.C., 1969.

Gordon, C. G., W. J. Galloway, D. L. Nelson, and B. A. Kugler. Evaluation of Highway Noise. Unpublished Report for Program NCHRP 3-7/1, Highway Research Board, National Academy of Sciences, Washington, D.C., January, 1970.

Harris, Cyril S., and Brian H. Aitken. "Noise in Subway Cars," *Sound and Vibration,* 5:2, February, 1971.

Horonjeff, R., and Soroka, W. W. *Transportation System Noise Generation, Propagation and Alleviation Phase I, Parts I and II.* Prepared for U.S. Department of Transportation, September, 1970.

Johnson, D. R., and F. G. Saunders, "The Evaluation of Noise from Freely Flowing Road Traffic," *Journal of Sound and Vibration,* 7:2, 1968, pp. 287–309.

Kibbee, Lewis C. *The Noise Trucks Make.* Washington, D.C.: American Trucking Association, Inc. September, 1964.

Millard, R. S. *A Review of Road Traffic Noise.* Road Research Laboratory, Crothorne, Berkshire, England, 1970.

Mills, C. H. G. "The Measurement of Traffic Noise," *The Control of Noise.* London: HMSO, 1962.

Norbye, Jan P., and Jim Dunne. "The 1970 Personal Cars Combine Luxury and Sportiness," *Popular Science,* 196:2, 1970.

Northwood, T. D. "Rail Vehicle Noise," Joint Railway Conference Paper No. 62-EIC-RRI, *Engineering Journal of the Engineer Institute of Canada,* 46, January, 1963, pp. 30–33.

Parsons, Brinckerhoff, Tudor, Bechtel. "San Francisco Bay Area Rapid Transit District Demonstration Project." Report No. 8, Acoustical Studies, June, 1968.

Pratt, Hugh W. *Noise and the North-East Expressway.* Prepared for the Northeast Citizens Planning Council of Philadelphia, June, 1967.

Purdy, Ken W. "The Mini Revolution," *Playboy,* 18:3, March, 1971.

Thiessen, George. "Community Noise—Surface Transportation," *Sound and Vibration,* 2:4, April, 1968.

——. "Community Noise Levels," in *Transportation Noises.* Ed. by J. D. Chalupnik. Seattle: University of Washington Press, 1970, pp. 23–32.

——. "Survey of the Traffic Noise Problem." Presented at the 69th Meeting of the Acoustical Society of America, Washington, D.C., June 2, 1965.

WATER
TRANSPORTATION

Anon. "Hovercraft for Quiet Ferrying," *Engineering*, 205:5313, 1968, p. 250.

Campbell, R. A. "A Survey of Passby Noise from Boats," *Sound and Vibration*, 3:9, 1969, pp. 24–26.

Cross, I., and C. O'Flaherty. "Introduction to Hovercraft and Hoverports," *Hovering Craft and Hydrofoil*, 9:6, 1970, pp. 18–28.

INDUSTRY

American Industrial Hygiene Association. *Industrial Noise Manual*, 2nd ed. Detroit, 1966.

Glorig, Aram. "Industrial Noise and the Worker," *Noise as a Public Health Hazard*. Washington, D.C.: American Speech and Hearing Association, 1969.

Goodfriend, Lewis. "Industry/Community Interaction on Noise Problems." Paper presented to Committee on Hearing, Bioacoustics and Biomechanics, National Academy of Science-National Research Council, April 1, 1971.

Hosey, A. D., and C. H. Powell, eds. *Industrial Noise, A Guide to Its Evaluation and Control*. U.S. Department of Health, Education and Welfare, Public Health Service Publication No. 1572. Washington, D.C.: GPO, 1967.

Morse, K. M. "Community Noise—The Industrial Aspect," *American Industrial Hygiene Association Journal*, 29:4, July–August, 1968, pp. 368–380.

Stichting Concawe. "A Guide to the Evaluation of Noise around Refineries." (The Hague, Netherlands.) Report prepared by Working Group Noise Control, May, 1968. Mimeographed.

OTHER

Anon. "US Bans 39 Toys under Safety Act," *New York Times*, December 23, 1970.

Baade, Peter K. "Household Noise Problems." Presented at the 80th Meeting Acoustical Society of America, Houston, Texas, November 3–6, 1970.

Bird, David. "Sirens Scream for Quiet's Sake," *New York Times*, December 19, 1969.

Carmody, Deirdra. "Blast of Construction Shatters Nerves and Windows," *New York Times*, September 26, 1970.

Consumers Union. "Bang, bang, you're deaf!" *Consumer Reports*, 35:11, November, 1970.

Consumers Union. "Table Saws," *Consumer Reports*, 35:5, May, 1970.

Gjaevenes, K. "Measurements on the Impulsive Noise from Crackers and Toy Firearms," *Journal of the Acoustical Society of America*, 39, 1966.

Hodge, David C., and R. Bruce McCommons. "Acoustical Hazards of Children's Toys," *Journal of the Acoustical Society of America*, 40:4, October, 1966.

Shuldiner, Herbert. "Here Come the 1971 Snowmobiles," *Popular Science*, 197:4, October, 1970.

U.S. Department of Commerce. *Snowmobiles*. Washington, D.C.: GPO, September, 1969.

U.S. Public Health Service. *Noise in Hospitals: An Acoustical Study of Noises Affecting the Patient*. Washington, D.C., 1963.

Ando, Y., and H. Hattori. "Effects of Intense Noise during Fetal Life Upon Postnatal Adaptability," *Journal of the Acoustical Society of America,* 74:4, 1970, pp. 1128–1130.

Anticaglia, J. "Extra-auditory Effects of Sound on the Special Senses," *Physiological Effects of Noise,* ed. by B. and A. Welch. New York: Plenum Press, 1970.

Bond, James. "Effects of Noise on the Physiology and Behavior of Farm Raised Animals." Paper presented at the AAAS Symposium, Boston, Mass., 28–30 December 1969.

Davis, H., and S. R. Silverman, eds. *Hearing and Deafness,* 3d ed. New York: Holt, Rinehart and Winston, 1970.

Davis, Hallowell, ed. *Auditory and Non-Auditory Effects of High Intensity Noise,* U.S. Naval School of Aviation Medicine, June, 1958.

Fink, Gregory B., and Iturrian, W. B., "Influence of Age, Auditory Conditioning, and Environmental Noise on Sound-induced and Seizure Threshold in Mice." *Physiological Effects of Noise,* ed. by B. and A. Welch. New York: Plenum Press, 1970.

Finkle, A. L., and Popper, J. R., "Clinical Effects of Noise and Mechanical Vibrations of a Turbo-jet Engine on Man," *Journal of Applied Physiology,* 1, 1948, p. 183.

Fletcher, Harvey. *Speech and Hearing in Communication.* New York: Van Nostrand Co., 1953.

Glorig, Aram. "Effects of Noise on Man," *Journal American Medical Association,* 196:10, June 6, 1966.

———. *Noise and Your Ear.* New York: Grune & Stratton, 1958.

Goromosov, M. S. *The Physiological Basis of Health Standards for Dwellings,* Public Health Paper #33. Geneva: World Health Organization, 1968.

Heinemann, Jack M. "Effect of Sonic Booms on the Hatchability of Chicken Eggs and Other Studies of Aircraft-generated Noise Effects on Animals," *Physiological Effects of Noise.* Ed. by B. and A. Welch. New York: Plenum Press, 1970.

Jansen, Gerd. "Effects of Noise on Physiological State," *Noise as a Public Health Hazard.* Washington, D.C.: American Speech and Hearing Association, 1969, pp. 89–99.

Krushinsky, L. N., *et al.* "The Functional State of the Brain during Sonic Stimulation," *Physiological Effects of Noise.* Ed. by B. and A. Welch. New York: Plenum Press, 1970.

Neher, G. M., *et al. The Role of Noise as a Physiologic Stressor.* Prepared under contract, U.S. Department of Health, Education and Welfare, 1969.

Rosen, Samuel. "Noise Pollution: A Need for Action," *Medical Tribune,* January 4, 1968.

Shatalov, N. N., and M. A. Murov. "Effect of Intensive Noise and Neuro-psychic Tension on the Level of Arterial Pressure and Extent of Hypertensive Vascular Disease," *Klinicheskaya Meditsina* (Clinical Medicine), Number 3, 1970, pp. 70–73.

Sontag, Lester, and A. Arvay. "Effect of Noise during Pregnancy upon

Fetal Viability and Development," *Physiological Effects of Noise,* Ed. by B. Welch. New York: Plenum Press, 1970.

Welch, Bruce L., and A. S. Welch. *Physiological Effects of Noise.* New York: Plenum Press, 1970.

HEARING LOSS

Baughn, William L. "Noise Control—Percent of Population Protected," *International Audiology,* 5:9, September, 1966.

Botsford, James H. "Scales for Expressing Noise Level-Damage Risk." Presented at the Symposium, Evaluating the Noises of Transportation, University of Washington, Seattle, Washington, March 26–28, 1969.

——. "Damage Risk," in *Transportation Noises* ed. by J. D. Chalupnik. Seattle, Washington: University of Washington Press, 1970.

Cohen, Alex, *et al.* "Sociocusis—Hearing Loss from Non-occupational Noise Exposure," *Sound and Vibration,* 4:11, November, 1970.

Consumers Union. "Bang bang, you're deaf!" *Consumer Reports,* 35:11, November, 1970.

Davis, Hallowell, and Richard S. Silverman. *Hearing and Deafness.* Revised edition. New York: Holt, Rinehart and Winston, 1962.

Dey, Frederick L. "Auditory Fatigue and Predicted Permanent Hearing Defects from Rock and Roll Music," *The New England Journal of Medicine,* 282:9, February 26, 1970.

Dougherty, J. D., and Welsh, O., "Community Noise and Hearing Loss," *New England Journal of Medicine,* 127:14, October 6, 1966.

Eldred, D. M., W. J. Gaoono, and H. von Gierke. "Criteria for Short-time Exposure to High Intensity Jet Aircraft Noise." WADC Technical Report No. 55-355, U.S. Air Force, Wright-Patterson Air Force Base, Ohio, 1955.

Glorig, A., W. D. Ward, and J. Nixon. "Damage Risk Criteria and Noise-Induced Hearing Loss," *Archives of Otolargyngology,* 74, 1961, p. 413.

Hodge, David C., and R. B. McCommons. "Acoustical Hazards of Children's Toys," *Journal of Acoustical Society of America,* 40:4, October, 1966.

International Standards Organization. "Assessment of Noise-Exposure During Work for Hearing Conservation Purposes." Draft proposal, ISO/TC 43WG8, June, 1968.

Keim, R. J. "Impulse Noise and Neorosensory Hearing Loss. Relationship to Small Arms Fire," *California Medicine,* 113:8, September, 1970, pp. 16–19.

Knudsen, V. O. "Noise, the Bane of Hearing," *Noise Control,* 1:11–13, May, 1955.

Kryter, K. D., W. D. Ward, J. E. Miller, and D. H. Eldredge. "Hazardous Exposure to Intermittent and Steady-state Noise," *Journal of Acoustical Society of America,* 39, 1966, pp. 451–464.

Kryter, K. D. "Evaluation of Exposures to Impulse Noise," *Archives of Environmental Health,* 20:5, May, 1970.

Poynor, R. E., and F. H. Bess. "The Effects of Snowmobile Engine Noise on Hearing." Paper presented at 46th Meeting, American Speech and Hearing Association, November, 1970.

Rosenwinkel, N., and K. Stewart. "Relationship of Hearing Loss to Steady-state Noise Exposure," *American Industrial Hygiene Quarterly,* Vol. 18, 1957.

Sataloff, Joseph. *Industrial Deafness: Hearing, Testing and Noise Measurement.* New York: McGraw-Hill, 1957.

———. "Temporary and Permanent Hearing Loss," *Archives of Environmental Health,* 10:1, January, 1965.

Sataloff, Joseph, and Zapp, John A., "The Environment in Relation to Otologic Disease," *Archives of Environmental Health,* 10:3, March, 1965.

Sataloff, Joseph, L. Vassallo, and H. Menduke. "Hearing Loss from Exposure to Interrupted Noise," *Archives of Environmental Health,* 18:6, June, 1969.

Shearer, William M. "Acoustic Threshold Shift from Power Lawnmower Noise," *Sound and Vibration,* 2:10, October, 1968.

U.S. Department of Health, Education and Welfare. "Characteristics of Persons with Impaired Hearing." PHS Publication #1000, Series 10, Number 35. Washington, D.C.: GPO, 1967.

Ward, W. Dixon. "The Effects of Noise on Hearing Thresholds," *Noise as a Public Health Hazard,* Washington, D.C.: American Speech and Hearing Association, 1969.

Ward, W. D. "Hearing Damage," in *Transportation Noises.* Ed. by J. D. Chalupnik. Seattle: University of Washington Press, 1970, pp. 174–186.

Ward, W. D., and D. Nelson. "Noise-Induced Permanent Threshold Shifts and the Equal Energy Hypothesis in the Chinchilla." Presented Acoustical Society of America, Atlantic City, New Jersey, April 21, 1970.

NEURAL-HUMORAL STRESS

Anthony A., and E. Ackerman. "Effects of Noise on the Blood Sosinophil and Adrenals of Mice," *Journal of the Acoustical Society of America,* 27:6, 1955, pp. 1144–1149.

Anthony A., E. Ackerman, and J. A. Lloyd. "Noise Stress on Laboratory Rodents. I: Behavioral and Endocrine Responses of Mice, Rats, and Guinea Pigs," *Journal of the Acoustical Society of America* 31:11, 1959, pp. 1430–1437.

Arguelles, M. A., *et al.* "Endocrine and Metabolic Effects of Noise in Normal, Hypertensize and Psychotic Patients," *Physiological Effects of Noise.* Ed. by B. and A. Welch. New York: Plenum Press, 1970.

Buckley, Joseph P., and Harold H. Smookler. "Cardiovascular and Biochemical Effects of Chronic Intermittent Neurogenic Stimulation," *Physiological Effects of Noise.* Ed. by B. Welch. New York: Plenum Press, 1970.

Foster, Francis M. "Human Studies of Epileptic Seizures Induced by Sound and Their Conditional Extinction," *Physiological Effects of Noise.* Ed. by B. and A. Welch. New York: Plenum Press, 1970.

Fuller, John L., and Robert C. Collins. "Genetic and Temporal Characteristics of Audiogenic Seizures in Mice," *Physiological Effects of Noise.* Ed. by B. and A. Welch. New York: Plenum Press, 1970.

Geber, William F. "Cardiovascular and Teratogenic Effects of Chronic Intermittent Noise Stress," *Physiological Effects of Noise.* Ed. by B. and A. Welch. New York: Plenum Press, 1970.

Henry, Kenneth R., and Robert Bowman. "Acoustics Priming of Audiogenic Seizures in Mice," *Physiological Effects of Noise.* Ed. by B. and A. Welch. New York: Plenum Press, 1970.

Jansen, Gerd. "Relation between Temporary Threshold Shift and Periphery

Circulatory Effects of Sound," *Physiological Effects of Noise.* Ed. by B. and A. Welch. New York: Plenum Press, 1970.

Jansen, G., *et al.* "Vegatative Reactions to Auditory Stimuli," *Transactions of the American Academy of Opthalmology and Otolaryngology,* May–June, 1964.

Jensen, Marcus M. and Pasmussen, "Audiogenic Stress and Susceptability to Infection," *Physiological Effects of Noise.* Ed. by B. and A. Welch. New York: Plenum Press, 1970.

Lockett, Mary F. "Effects of Sound on Endocrine Function and Electrolyte Excretion," *Physiological Effects of Noise.* Ed. by B. and A. Welch. New York: Plenum Press, 1970.

Lukas, J. S., Donald J. Peeler, and Karl D. Kryter. *Effects of Sonic Booms and Subsonic Jet Flyover Noise on Skeletal Muscle Tension and a Placed Tracing Task.* National Aeronautics and Space Administration, Washington, D.C., February, 1970.

Niemeyer, W. "Speech Discrimination in Noise Induced Deafness," *International Audiology,* 6:6, June, 1957.

Rosen, Samuel. "Noise, Hearing and Cardiovascular Function," *Physiological Effects of Noise.* Ed. by B. and A. Welch. New York: Plenum Press, 1970.

Shetalow, N. N., A. D. Sartausy, and K. V. Giotova. "On the State of the Cardiovascular System under Conditions of Exposure to Continuous Noise," *Labor Hygiene and Occupational Diseases,* No. 6, 1962, pp. 10–14.

Svadlrouskaya, N. F. "Effect of Long-Term Noise on Cerebral Oxidation Processes in Albino Rats" (Russian), *Gigiena I. Sanitaria,* Volume 7, 1967.

Tamari, I. "Audiogenic Stimulation and Reproductive Function," *Physiological Effects of Noise.* Ed. by B. and A. Welch. New York: Plenum Press, 1970.

SLEEP

Corcoran, D. W. J. "Shorter Articles and Notes: Noise and Loss of Sleep," *Quarterly Journal of Experimental Psychology,* 14:8, August, 1965.

Dement, William C. "Recent Studies on the Biological Role of Rapid Eye Movement Sleep," *The American Journal of Psychiatry,* 22:10, October, 1965.

——. "The Effect of Partial REM Sleep and Delay Recovery," *Journal of Psychiatric Research,* 4:2, April, 1966.

Ephron, H. S., and P. Carrington. "Rapid Eye Movements and Cortical Homeostasis," *Psychological Review,* 73:6, 1966, pp. 500–526.

Ferguson, J., and W. Dement. "The Effect of Variations in Total Sleep Time on the Occurrence of Rapid Eye Movement Sleep in Cats," *Electroencephology and Clinical Neurophysiology,* 22, July, 1967.

Johnson, Laverne. "Auditory Response Thresholds and Habituation During Sleep." Paper presented to Committee on Hearing, Bioacoustics and Biomechanics, National Academy of Science-National Research Council, March 31, 1971.

Jouvet, M. "Report on Research Progress." Paper read at the Association for the Psychophysiological Study of Sleep, Washington, D.C., March, 1965.

Kryter, Karl. "Effects of Aircraft Noise and Booms on Sleep of Different Age Groups." Paper presented to Committee on Hearing, Bioacoustics and Biomechanics, National Academy of Science-National Research Council, March 31, 1971.

Kramer, Milton. "Noise-Disturbed Sleep and Post-Sleep Performance."

Paper presented to Committee on Hearing, Bioacoustics and Biomechanics, National Academy of Science-National Research Council, March 31, 1971.

Luce, G. G., and Dennis McGinity. *Current Research in Sleep and Dreams.* U.S. Public Health Service, Department of Health, Education, and Welfare, Report 1389. Washington, D.C.: GPO, 1965.

Lukas, Jerome S., and Karl Kryter. *A Preliminary Study of the Awakening and Startle Effects of Simulated Sonic Booms.* Stanford Research Institute, April, 1968.

——. "Awakening Effects of Simulated Sonic Booms and Subsonic Aircraft," *Physiological Effects of Noise.* Ed. by B. and A. Welch. New York: Plenum Press, 1970.

Rechtschaffen, Allan. "Sleep, Dreams and Responses to Auditory Stimulation." Paper presented to Committee on Hearing, Bioacoustics and Biomechanics, National Academy of Science-National Research Council, March 31, 1971.

Thiessen, George J. "Effects of Noise during Sleep," *Physiological Effects of Noise.* Ed. by B. Welch. New York: Plenum Press, 1970.

Psycho-Social Effects
GENERAL

Anthony A., and J. E. Harclerode. "Noise Stress in Laboratory Rodents. II: Effects of Chronic Noise Exposures on Sexual Performance and Reproductive Function of Guinea Pigs," *Journal of the Acoustical Society of America,* 31:11, 1959, pp. 1437–1440.

Baron, Robert A. "The Noise Receiver: The Citizen," *Sound and Vibration,* 2:15, May, 1968.

Botsford, James H. "Using Sound Levels to Gauge Human Response to Noise," *Sound and Vibration,* 3:10, October, 1969.

Bragdon, C. R. "Community Noise and the Public Interest," *Sound and Vibration,* 3:12, December, 1969, pp. 16–21.

Broadbent, Donald. "Effects of Noises of High and Low Frequency on Behavior," *Ergonomics,* 1:1, November, 1967.

——. "Effects of Noise on Behavior," *Handbook of Noise Control.* Ed. by Cyril Harris. New York: McGraw-Hill, 1957.

Brodey, Warren. "Sound and Space," *Journal American Institute of Architects,* 42:7, July, 1964.

Cohen, Alexander. "Noise and the Psychological State," *Noise as a Public Health Hazard,* Washington, D.C.: American Speech and Hearing Association, 1969.

Corso, John F. *The Effects of Noise on Human Behavior.* Wright Air Development Center: WADC Technical Report, 53-81, 1952.

Culbert, S., and M. Posner. "Human Habituation to an Acoustical Energy Distribution Pattern," *Journal of Applied Psychology,* 44, 1966, pp. 263–266.

Doniger, S., and M. DeGossely. "Modification of the Sensitivity to Noise in the Course of Autogenic Training: Clinical Study and EEG," *Psychotherapy and Psychosomatics,* 15:1, 1967.

Eldredge, D., *et al.* "Proactive Inhibition Recency and Limited-Channel Capacity under Acoustic Stress," *Perceptual Motor Skills,* August, 1967.

Finkleman, J., and D. Glass. "Reappraisal of the Relationship between Noise and Human Performance by Means of a Subsidiary Task Measure," *Journal of Applied Psychology,* 54:3, June, 1970, pp. 211–213.

Freed, M. I. "Our Deaf Employees Are Not Handicapped," *Rehabilitation Record,* 3:3, May–June, 1962, pp. 34–36.

Glass, David, *et al.* "The Psychic Cost of Adaptation to an Environmental Stressor," November, 1968. Pre-publication draft.

Goldfarb, W., and Mintz, J., "The Schizophrenic Child's Reaction to Time and Space," *Archives of General Psychiatry,* 5, 1961.

Grandjean, E. "Physiological and Psychological Effects of Noise," *Menschan Limmelt,* 1960.

Gulian, E. "Effects of Noise on an Auditory Vigilance Task," *Revue Romaine des Sciences Sociales: Series de Psychologie,* 10:2, 1966, p. 175.

Hall, Edward T. *The Hidden Dimension.* New York: Doubleday, 1966.

——. "Proxemics: The Study of Man's Spatial Relations," *Man's Image in Medicine and Anthropology.* New York: International Universities Press, 1963.

Harris, C. S. *The Effects of High Intensity Noise on Human Performance.* AMRL Technical Report No. 67-119 iii, U.S. Air Force, 1968.

Hoffman, J. E. "The Effect of Noise on Intellectual Performance as Related to Personality and Social Factors in Upper Division High School Students," *Dissertation Abstracts,* 27:6-A, 1966, 1658.

Hormann, H., and U. Osterkamp. "On the Influence of Intermittent Noise upon the Organization of Memory Content," *Zeitschrift for Experimentelle und Angewandte Psychologie,* 13:2, 1966, pp. 265–278.

Horowitz, M. J. "Spatial Behavior and Psychopathology," *Journal of Nervous and Mental Disease,* 140:1, January, 1968.

Isumiyama, M. "Studies of Influences of Train Noise upon School Children: II. On Speech Articulation," *Tohoku Psychologica Folia,* 23:1–2, 1964.

Lindsey, Robert. "Irate Citizens across the Nation Are Vigorously Resisting the Construction of Jetports," *New York Times,* December 22, 1969.

McCartney, J. L. "Noise Drives Us Crazy." Reprinted by the National Noise Abatement Council, New York City, from *Pennsylvania Medical Journal,* August, 1941.

Mapes, Glynn. "A Vacuum's Woosh, A Car Door's Thunk Don't Just Happen," *Wall Street Journal,* September 10, 1968.

Maruyama, K. "Studies of Influence of Train Noise Upon School Children: VI: GSR During Mental Calculation," *Tohoku Psychologica Folia,* 23:1–2, 1964.

Michalova, C., and V. Hrubes. "Frequency of Neuroses and Psychosomatic Diseases Mainly Hypertension under the Influence of Noise." Unpublished Report.

Mott, Richard H., *et al.* "Comparative Study of Hallucinations," *Archives of General Psychiatry,* 12:6, June, 1965, and Ludwig, Arnold M., "Auditory Studies in Schizophrenia," *American Journal of Psychiatry,* 119:8, August, 1962.

Nicholi, Armand M. "The Motorcycle Syndrome," *American Journal of Psychiatry,* 126:11, May, 1970, pp. 1588–1591.

Pollack, K. G., and F. C. Bartlett. *Psychological Experiments on the Effects of Noise.* Report of the Industrial Research Board, No. 65. London, 1932.

Vernon, McCay. "Sociological and Psychological Factors Associated with Hearing Loss," *Journal of Speech and Hearing Research,* 12:3, September, 1969.

Wydon, D. P. "Studies of Children under Imposed Noise and Heat Stress," *Ergonomics,* 13:9, September, 1970.

Webster, John C., and M. Lepor. "Noise You Can Get Used to It," *Journal of the Acoustical Society of America,* 45:3, March, 1969.

Anon. "Driver Accident Rate—Hearing vs. Deaf," *New York Times,* March 19, 1968, and "Deaf People Called Excellent Drivers," *New York Times,* February 18, 1970.

Auble, D., and N. Britton. "Anxiety as a Factor Influencing Performance under Auditory Stimulation," *Journal of General Psychology,* 58, 1958, pp. 111–114.

Boggs, D. H., *et al.* "Differential Effect of Noise on Tasks of Varying Complexity," *Journal of Applied Psychology,* S2, April, 1968, pp. 148–153.

Broadbent, D. E., and E. A. Litte. "Effects of Noise Reduction in a Work Situation," *Occupational Psychology,* 34, 1960, pp. 133–140.

Burg, F. D., *et al.* "Licensing the Deaf Driver," *Archives of Otolargyngology,* 91:3, March, 1970.

Carpenter, A. "Effects of Noise on Performance and Productivity," *The Control of Noise.* London: HMSO, 1962.

Cassel, E., and K. Dallenbach. "The Effect of Auditory Stimulation upon the Sensory Reaction," *American Journal of Psychology,* 1918, pp. 129–143.

Hamilton, P., *et al.* "The Effect of Alcohol and Noise on Components of a Tracking and Monitoring Task," *British Journal of Psychology,* 61:5, May, 1970.

Doniger, S., and M. DeGossely. "Modification of the Sensitivity to Noise in the Course of Autogenic Training: Clinical Study and EEG," *Psychotherapy and Psychosomatics,* 15:1, 1967.

Eldredge, D., *et al.* "Proactive Inhibition Recency and Limited-Channel Capacity under Acoustic Stress," *Perceptual Motor Skills,* August, 1967.

Finkleman, J., and D. Glass. "Reappraisal of the Relationship between Noise and Human Performance by Means of a Subsidiary Task Measure," *Journal of Applied Psychology,* 54:3, June, 1970, pp. 211–213.

Freed, M. I. "Our Deaf Employees Are Not Handicapped," *Rehabilation Record,* 3:3, May–June, 1962, pp. 34–36.

Gulian, E. "Effects of Noise on an Auditory Vigilance Task," *Revue Romaine des Sciences Sociales: Series de Psychologie,* 10:2, 1966, p. 175.

Harris, C. S. *The Effects of High Intensity Noise on Human Performance.* AMRL Technical Report No. 67-119 iii, U.S. Air Force, 1968.

Hoffman, J. E. "The Effect of Noise on Intellectual Performance as Related to Personality and Social Factors in Upper Division High School Students," *Dissertation Abstracts,* 27:6-A, 1966, 1658.

Hormann, H., and U. Osterkamp. "On the Influence of Intermittent Noise upon the Organization of Memory Content," *Zeitschrift for Experimentelle und Angewandte Psychologie,* 13:2, 1966, pp. 265–278.

Hovey, H. B. "Effects of General Distraction on the Higher Thought Processes," *American Journal of Psychology,* 40, 1928, pp. 585–591.

Jerison, H., and S. Wing. "Effects of Noise and Fatigue on a Complex Vigilance Task." WADC Technical Report No. 57-114, U.S. Air Force, Wright-Patterson Air Force Base, Ohio, 1957.

———. "Performance on a Simple Vigilance Task in Noise and Quiet," *Journal of the Acoustical Society of America,* 29, 1957, pp. 1163–1165.

HUMAN
PERFORMANCE

Laird, D. "The Influence of Noise on Production and Fatigue as Related to Pitch, Sensation Level, and Steadiness of Noise," *Journal of Applied Psychology,* 17, 1933, pp. 320–330.

Lehman, D., *et al.* "An Investigation of the Effects of Various Noise Levels as Measured by Psychological Performance and Energy Expenditure," *Journal of School Health,* 35:3, March, 1965.

Loeb, M. *The Influence of Intense Noise on Performance of a Precise Fatiguing Task.* U.S.A. Medical Research Laboratory Report No. 268 ii, 1957.

Luckiesh, M. "Visual Efficiency in Quiet and Noisy Work Places," *Electronics World,* 98, 1931, pp. 472–473.

Pickett, J. M. "Message Constraints, A Neglected Factor in Predicting Industrial Speech Communication," *Noise as a Public Health Hazard.* Washington, D.C.: American Speech and Hearing Association, 1969.

Slater, Barbara R. "Effects of Noise on Pupil Performance, *Journal of Educational Psychology,* 59:4, August, 1968.

Stevens, S. S. *The Effects of Noise and Vibrations on Psychomotor Efficiency.* OSRD Report No. 32, Psycho-acoustics Laboratory, Harvard University, 1941.

Takahashi, I., and S. Kyo. "Studies on the Differences of Adaptabilities to the Noisy Environment in Sexes and the Growing Environment," *Journal of the Anthropological Society of Nippon,* 76:758, 1968, pp. 24–51.

Viteles, M., and K. Smith. "An Experimental Investigation of the Effects of Change in Atmospheric Conditions and Noise upon Performance," *Transactions of the American Society of Heating and Ventilating Engineers,* 52:1291, 1946, pp. 167–182.

Vontess, C. E. "The Effects of Stressor Conditions on Test Scores of a Selected Group of High School Students," *Dissertation Abstracts,* 27:6-A, 1966, p. 1670.

Webster, J. C. "The SIL—Past, Present and Future," *Sound and Vibration,* 3:8, August, 1969.

Webster, J., and R. Klumpp. "Effects of Ambient Noise and Nearby Talkers on a Face-to-Face Communication Task," *Journal of the Acoustical Society of America,* 34:7, 1962.

Woodhead, M. "Effect of Brief Loud Noise on Decision Making," *Journal of the Acoustical Society of America,* 31, 1959, pp. 1329–1331.

——. "The Effect of Bursts of Noise on an Arithmatic Task," *American Journal of Psychology,* 77:4, 1964, pp. 627–633.

HUMAN RESPONSE

Borsky, P. N. *Community Reactions to Air Force Noise. I: Basic Concepts and Preliminary Methodology; II: Data on Community Studies and Their Interpretation.* WADD Technical Report 60-689, U.S. Air Force, Wright-Patterson Air Force Base, Ohio, March, 1961.

——. *Community Reactions to Noise in the Oklahoma City Area.* Technical Report, 65-37, Aerospace Medical Research Laboratories, February, 1965.

——. "Effects of Noise on Community Behavior," in *Noise as a Public Health Hazard.* Ed. by W. D. Ward and J. E. Fricke. Washington, D.C.: American Speech and Hearing Association, 1969, pp. 187–201.

Cedarlof, H., E. Jonsson, and A. Kajland. "Annoyance Reactions to Noise from Motor Vehicles: An Experiment Study," *Acustica,* 13:4, 1963, pp. 270–279.

Clark, W. *Reaction to Aircraft Noise.* ASD Technical Report 61-610, U.S. Air Force, November, 1961.

Franken, P. A., and G. Jones. "On Response to Community Noise," *Applied Acoustics,* 2:4, October, 1969.

Fukushima, K. "Aircraft Acoustics: Community Noise Prediction." The Boeing Co., Renton, Washington, Report No. D6-4219TN (July 2, 1968). N68-32202.

Griffiths, I. D., and F. J. Langdon. "Subjective Response to Road Traffic Noise," *Building Research Current Papers,* 37/38, April, 1968, pp. 16–32.

Hazard, W. R. "Community Reactions to Aircraft Noise, Public Reactions," in National Aeronautics and Space Administration SP-189, October, 1968, pp. 661–672.

Hecker, M. H. L., and K. D. Kryter. *Comparisons between Subjective Ratings of Aircraft Noise and Various Objective Measures.* Prepared for Federal Aviation Administration, Report No. 68-33, April, 1968.

Hess, John L. "French Towns Sue over Airport Noise," *New York Times,* July 11, 1970.

Klausner, Samuel Z. "Noise Annoyance: Some Social Psychological Implications of a Physical Event." Panel on Noise Abatement, U.S. Department of Commerce Advisory Board, January, 1969. Mimeographed.

Kryter, K. D. "Psychological Reactions to Aircraft Noise," *Science,* 153, March 18, 1966.

Landgon, F. J., and W. E. Scoles. *The Traffic Noise Index: A Method of Controlling Noise Nuisance. Building Research Station Current Papers* 38/68, April, 1968.

Lindsey, Robert. "Irate Citizens across the Nation Are Vigorously Resisting the Construction of Jetports," *New York Times,* December 26, 1969.

Lundberg, B. "The Acceptable Nominal Sonic Boom Overpressure in SST Operation," *Noise as a Public Health Hazard,* Washington, D.C.: American Speech and Hearing Association, 1969.

McKennel, A. C. "Complaints and Community Action," in *Transportation Noises.* Ed. by J. D. Chalupnik. Seattle: University of Washington Press, 1970, pp. 228–244.

Nixon, C. W. "Human Response to Sonic Booms," *Aerospace Medicine,* 36, 1965, p. 399.

Nixon, C. W., and H. H. Hubbard. *Results of the US Air Force-National Aeronautics and Space Administration Flight Program to Study Community Responses to Sonic Booms in the Greater St. Louis Area.* National Aeronautics and Space Administration TN 2075, May, 1965.

Robinson, D. W. *The Concept of Noise Pollution Level.* National Physical Laboratory, Aerodynamics Division (England) NPL Aero Report Ac 38, March, 1969.

Smith, H. K. "Jet Planes Hurt Hearing Bring Protest to Capitol," *Philadelphia Bulletin,* March 26, 1967.

Stevens, K. N., W. A. Rosenblith, and R. H. Bolt. "A Community's Reaction to Noise: Can It Be Forecast?" *Noise Control,* 1, January, 1955, pp. 63–71.

Well, R. J. "Jury Ratings of Complex Aircraft Noise Spectra Versus Calculated Ratings." Presented at the 80th Meeting Acoustical Society of America, Houston, Texas, November 3–6, 1970.

Index

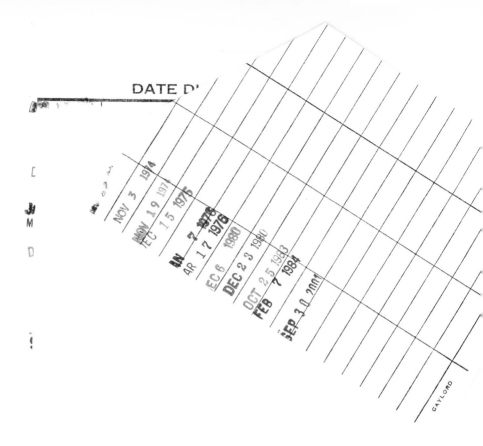